The Future Computed

ARTIFICIAL INTELLIGENCE AND ITS ROLE IN SOCIETY

（中英文对照）

计算未来

人工智能及其社会角色

沈向洋（Harry Shum） 〔美〕施博德（Brad Smith） ■ 编著

北京大学出版社
PEKING UNIVERSITY PRESS

图书在版编目(CIP)数据

计算未来:人工智能及其社会角色/沈向洋,(美)施博德编著.—北京:北京大学出版社,2018.9
ISBN 978-7-301-29831-2

Ⅰ.①计⋯　Ⅱ.①沈⋯　②施⋯　Ⅲ.①人工智能—研究　Ⅳ.①TP18

中国版本图书馆 CIP 数据核字(2018)第 189506 号

书　　　名	计算未来——人工智能及其社会角色 JISUAN WEILAI——RENGONG ZHINENG JIQI SHEHUI JUESE
著作责任者	沈向洋　〔美〕施博德　编著
责任编辑	杨玉洁　陈晓洁
标准书号	ISBN 978-7-301-29831-2
出版发行	北京大学出版社
地　　　址	北京市海淀区成府路 205 号　100871
网　　　址	http://www.pup.cn　http://www.yandayuanzhao.com
电子信箱	yandayuanzhao@163.com
新浪微博	@北京大学出版社　@北大出版社燕大元照法律图书
电　　　话	邮购部 62752015　发行部 62750672　编辑部 62117788
印　刷　者	北京中科印刷有限公司
经　销　者	新华书店
	880 毫米×1230 毫米　32 开本　9.375 印张　194 千字 2018 年 9 月第 1 版　2019 年 5 月第 4 次印刷
定　　　价	49.00 元

未经许可,不得以任何方式复制或抄袭本书之部分或全部内容。
版权所有,侵权必究
举报电话:010-62752024　电子信箱:fd@pup.pku.edu.cn
图书如有印装质量问题,请与出版部联系,电话:010-62756370

目录

第1章
人工智能的未来 — 1

微软对人工智能的态度 — 8
现代人工智能的潜力——应对社会挑战 — 18
人工智能带来的挑战 — 22

第2章
负责任地使用人工智能的原则、政策和法律 — 23

伦理和社会影响 — 26
制定人工智能相关的政策和法律 — 41
促进对话和分享最佳实践 — 49

第3章
人工智能与职业和工作的未来 — 51

技术对职业和工作的影响 — 56
工作、工作场所和职业的性质不断变化 — 63
帮助每个人应对未来的工作挑战 — 70

改变规范，适应劳动者不断变化的需求　　83
通力合作　　92

结论
人工智能增强人类创造力　　93

补记
人工智能的中国使命　　99

数字经济——全球化普惠机遇　　104
智在中国，惠及全球　　106
能力有多大，责任就有多大　　109
有中国特色的 AI 之路　　110
总结　　118

注释　　119

第1章
人工智能的未来

*19*56年夏天，达特茅斯学院举行了一场特殊的会议。与会研究人员探讨的主题是，如何才能开发出像人类一样，有能力从经验中自主学习的计算机系统。这场会议标志着人工智能开发的纪元时刻。尽管如此，对机器智能概念的探索早在此之前已经开始，到达特茅斯会议召开时，已有十余年的历史。其中最著名的即为阿兰·图灵提出的图灵测试：如果人类在与一台机器进行交互时（在当年，交互的方式限于文本），无法辨别对方是人还是计算机，那么，这台机器就可以视为是"智能的"。

自达特茅斯会议召开以来，研究人员仍在数十年如一日地推进人工智能技术的进步。机器视觉、自然语言理解、推理、规划，和机器人学等分支学科的发展，带来了源源不断的创新，其中许多已经成为我们日常生活的一部分。导航系统的路线规划功能、从互联网海量信息中检索并排列内容的搜索引擎，以及邮递服务中使用的自动识别手写地址并将物件送达的机器视觉功能，无一不是通过人工智能来实现的。

微软认为，人工智能是赋予计算机感知、学习、推理及协助决策的能力，从而通过与人类相似的方式来解决问题的一组技术。在过去，计算机只能按照预先编写的固定程序开展工作，而具备该等能力以后，计算机理解世界以及与世界交互的方式，将比以前大为自然和灵敏。

就在不久以前，我们还只能通过命令行界面与计算机进行交互。图形用户界面向前迈出了重要的一步，而在不久的将来，我们将可以直接通过对话与计算机进行日常交互，就像与人交流一样。实际上，为了实现此类新功能，我们正在教计算机观看、倾听、理解和推理。关键技术包括：

视觉：计算机通过识别图片或视频中的内容来"看"的能力。

语音：计算机通过理解人们所说的话来"听"并将其转录成文字的能力。

语言：计算机把握语言中的诸多微妙差异和复杂性（例如俚语和惯用语），"理解"话语含义的能力。

知识：计算机通过理解人、事物、地点、事件等之间的关系来进行"推理"的能力。例如，当我们搜索某部电影时，便会得到关于演员阵容以及该等演员参演的其他电影的信息，或者在工作中，在开会的时候，你会自动收到最近曾和对方分享的几份文件。这些都是计算机通过推理从而就信息得出相关结论的例子。

计算机正在像人一样学习,即通过经验学习。计算机的经验是以数据的形式获得的。例如,计算机会结合一天中的时段、季节变化、天气状况以及该区域的重大事件(例如音乐会或体育赛事),再根据历史交通流量数据,来对交通状况进行预测。从更广泛的角度而言,丰富的信息"图谱"是计算机理解人、实体及事件之间的相关关系和交互的基础。在开发人工智能系统时,微软利用了多个信息图谱,其中包括关于世界、工作以及人的知识。

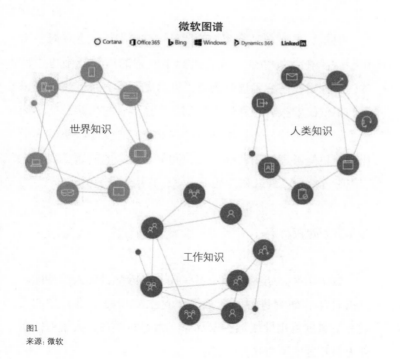

图1
来源:微软

过去几年，由于可用数据的大幅增加及其他原因，研究人员在这些技术上取得了重大进展。2015年，微软的研究人员使用标准ImageNet 1K图像数据库进行了一项测试，并宣布，他们已经教会计算机识别照片或视频中的客体，其准确度与人无异。[5] 2017年，微软的研究人员宣布，他们已经开发出一个语音识别系统，能够像专业的转录团队一样准确地理解口语；使用标准Switchboard数据集进行的测试显示，出错率仅为5.1%。[6]实质上，人工智能增强型计算机在大多数情况下可以像人一样准确地看和听。

但是，要把这些创新应用到日常生活中，还有很多工作要做。在背景嘈杂，或者碰到不熟悉的口音或陌生语言时，计算机在理解言语时仍有难度。尤其具有挑战性的是，我们不仅仅要教会计算机识别词语，还要让它理解那些话的意思、得出结论、进行推理，并在此基础上作出决策。为了让计算机能够理解语言的意义，回答更复杂的问题，我们需要把眼界放宽，理解和评估上下文，并引入背景知识。

为什么说时机已至？

数十年来，研究人员一直在孜孜不倦地研究人工智能。过去几年，研究进展不断加快，这主要得益于三方面的发展——数据可用性增加；云计算能力不断增强；人工智能研究人员开发出了更强大的算法。

随着我们生活的数字化程度越来越高,以及传感器的廉价和普及,计算机可用于学习的数据已经达到了空前的规模。

图2
来源:《2014年IDC数字宇宙预测》

有了数据,计算机才能识别模式(往往很微妙),从而去"看"、去"听"、去"理解"。

要对所有数据进行分析,需要有巨大的计算力,而云计算的高效恰恰为我们提供了这个能力。如今,任何类型的组织都能借用云计算的能力来开发和运行其人工智能系统。

微软、其他技术公司、大学以及政府的研究人员通过结合数据可用性和强大的计算力,实现了人工智能技术的突破——例如使用所谓的"深度神经网络"进行的"深度学习"——让计算机模仿人类的学习路径。

在很多方面，人工智能作为一种技术还远远没有成熟。迄今为止的大部分进展，都在于教会计算机执行指定的任务——玩游戏、识别图像、预测交通状况。要让计算机具备"通用"智能，还有很长的路要走。在通过触觉、视觉和嗅觉等感官来理解世界并与世界进行交互的能力等方面，现在的人工智能还不如孩童。对于人类表情、语气、情感以及人类交往的微妙之处，人工智能系统只有最初级的理解能力。换言之，如今的人工智能"IQ"很强，但"EQ"很弱。

微软正在努力赋予计算机更为精微的功能。我们相信，采用综合性的方法，将人工智能各个学科结合起来，将开发出更加精良的工具，帮助人们执行更为复杂、更多面性的任务。然后，随着我们学会如何将人工智能的多种IQ功能和人类与生俱来的能力结合起来——例如将从某项任务获得的知识应用于另一项任务、对世界的常识性理解、自然互动的能力，或者知道和区分别人的意图是搞笑还是讽刺的能力——人工智能的潜能将进一步发挥。这显然是一个艰巨的挑战，但是，当机器在交互过程中能够将IQ（智商）与EQ（情商）融为一体时，我们将实现我们所谓的"对话式AI"。这将是人机交互进化的重要一步。

微软对人工智能的态度

40年前，比尔·盖茨和保罗·艾伦成立微软时，他们的目标是让每一个人都能受益于计算机的能力——这种计算能力当时主要局限于大型机。他们着手建立用于家庭、学校以及工作场所，帮助人们提高效率的"个人"计算机。如今，微软通过人工智能所做的事情也与此类似。我们的目标是部署人人可用、体现永恒社会价值的人工智能系统，让人工智能惠及每一个人，也让人工智能赢得所有人的信任。[7]

增强人类创造力

我们相信，人工智能将创造无限广阔的机会，推动经济和社会全面进步。趋利避害的关键在于采用以人为本的方式开发人工智能。简言之，我们开发人工智能是为了增强人类的能力，尤其是人类天生的创造力。我们希望将计算机的能力与人类的能力相结合，帮助人类取得更大成就。

计算机非常善于记忆。只要系统没有发生故障，计算机就永远不会遗忘。计算机的概率推理能力也非常出色，而这是很多人的短板。计算机还非常善于识别对人们来说过于微妙、难以发现的数据模式。凭借这些能力，计算机能够帮助我们更好地作出决策。这是一个实实在在的好处，因为，正如认知心理学的研究人员所证实的那样，人类的决定往往并不完美。广义上讲，任何领域，只要智能本身在其中能够发挥作用，几乎都会因计算机提供的"计算智能"而受到重大影响。

• 人工智能帮助临床医生改进医学图像分析

实践中,人们已经开始应用人工智能系统来解决重大问题。InnerEye便是一个很好的例子。InnerEye项目是微软公司的英国研究人员与多位肿瘤医生联合开发的人工智能系统,目的是帮助医生更有效地治疗癌症。[8]

InnerEye采用的人工智能技术最初是为视频游戏开发的;它通过计算机断层扫描(CT)和磁共振成像(MRI)分析,帮助肿瘤医生更快地锁定癌症治疗的靶点。借助CT和MRI扫描,医生可以从三维角度查看病人身体内部情形,并研究肿瘤等身体异常情况。对于正在接受放射治疗的癌症患者，肿瘤医生可通过CT与MRI扫描，来划定肿瘤与周围健康组织、骨骼和器官的界线。这有助于将放射治疗集中在肿瘤上，尽可能避免对健康组织的细胞造成破坏。目前，这项3-D分界工作由医生手动操作，不但速度缓慢，且极易出错。放射肿瘤医生需要手工、逐一绘制数百个横断面图像的轮廓——整个过程需要耗费数小时之久。InnerEye的设计目的是在短时间内完成同样的工作，使肿瘤医生能够精准掌控最终划定的界线。

为创建InnerEye的自动分界功能，研究人员使用了成千上万份原始CT和MRI扫描图像（所有患者身份信息均已抹去）。这些图像被输入人工智能系统，供其学习识别、区分肿瘤和健康的解剖结构，并最终达到临床所需的准确度水平。一旦InnerEye完成自动分界，肿瘤医生便可着手对轮廓进行微调。一切都在医生的掌控之中。在功能进一步强化之后，InnerEye或许还可以用于测量和追踪肿瘤的变化情况，甚至评估治疗的效果。

- **人工智能帮助研究人员预防疾病爆发**

另一个值得一谈的例子是征兆计划（Project Premonition）。近年来，动物和昆虫传播的寨卡、埃博拉及登革热等高危疾病屡屡爆发，夺去了许多人的生命。流行病学家往往要在疫情爆发后，才能意识到这些病原体的存在。征兆计划由微软

研究院、匹兹堡大学、加州大学河滨分校和范德堡大学的科学家和工程师共同开发,目的是检测出环境中的病原体,以便公共卫生官员在疫情爆发之前为人们提供保护措施,预防疾病传播。[9]

流行病学家需要通过传感器来检测出现的病原体。该项目的研究人员提出了一个天才的想法:为什么不使用蚊子作为传感器呢?它们数量巨大,吸食的动物范围广泛,虽然只是吸取少量的血液,但已足以从中获取被叮咬动物的基因信息和环境中传播的病原体。

研究人员使用能够在复杂环境中飞行的先进无人机寻找蚊子滋生的地区,并在这些地区部署机械捕蚊器。这些捕蚊器能够根据翅膀的运动模式,将研究人员希望收集的蚊种与其他昆虫区分开来。收集到标本后,再通过云级别的基因组学和先进的人工智能系统,来识别蚊子吸食的动物以及这些动物携带的病原体。在过去,这种基因分析可能需要耗时一个多月,而现在,征兆计划的人工智能功能仅需大概12小时便能完成这一工作。

2016年寨卡疫情爆发期间,征兆计划在休斯敦对无人机和捕蚊器进行了测试。他们收集了9个品种共2万多只蚊子,包括已知携带寨卡病毒、登革热病毒、西尼罗河病毒和疟疾的蚊种。捕虫器在捕获昆虫的同时也收集环境状况数据,所以,这项测试不仅提供了有关环境中病原体的有用数据,还提供了有关蚊子习性的有用数据。因此,该项目的研究人员能够更容易地锁定蚊子滋生的重点区域。目前,研究人员正在努力作出改进,试图在识别已知疾病之外,检测出是否存在未知的病原体。

虽然该项目仍处于初期阶段,但它指明了一个方向。一旦建立成熟、有效的预警系统,便能够事先从环境中检测出某些世界上最危险的疾病,预防致命性疫情的爆发。

让以人为本的人工智能惠及大众

除非向大众普及人工智能,否则,我们无法兑现我们作出的人工智能的承诺。全世界的人们都可以受益于人工智能——但前提是人工智能技术得以足够普及。微软对这一承诺的履行，始于基础研发。微软研究院拥有26年历史，是全球领先的研究机构之一，始终致力于推动计算机科学的发展以及微软产品和服务的研发。我们的研究人员发表了逾22 000篇论文，内容涵盖各个研究领域，从环境到健康，从隐私到安全。不久前，我们宣布成立微软人工智能与研究事业部，这个全新的团队聚集了大约7 500名计算机科学家、研究人员和工程师，致力于深入了解智能的计算基础，专注于全面整合所有人工智能研究领域的研究，来解决人工智能领域一些最具挑战的问题。

我们始终鼓励研究人员广泛发表研究成果，以便世界各地的人工智能研究人员，包括大学、公司和政府机构等，能够充分利用这些成果。

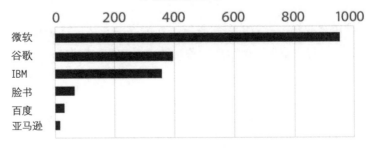

图3
来源:《经济学人》

我们正在逐步将各种人工智能功能融入最受客户欢迎的产品中，例如Windows和Office。人工智能系统能够检测恶意软件并自动保护计算机，因此提高了Windows的安全性。就Office而言，Researcher for Word可以帮助您撰写更为出色的文件。您可以在撰写文件的过程中，直接使用必应搜索的"知识图谱"查找并整合来自整个网络的相关信息，而不必另行打开浏览器。如果您想创建PowerPoint演示文稿，PowerPoint Designer会评估您使用的图像和文本，提供设计小贴士，提升幻灯片外观的专业性，为图像提供文本标题建议，使其更加便于理解。PowerPoint演示翻译插件可自动提供60多个语种的自动字幕，从而打破语言障碍，帮助您与观众进行更有效的互动。这一功能也是听障人士的福音。

微软个人数字助理小娜（Cortana）同样采用人工智能作为支持技术。小娜虽上线不久，但学习速度极快，已经可以帮您安排会议、预订餐厅，并就各种主题寻找问题的答案。假以时日，小娜将与其他个人数字助理实现交互，自动处理耗时耗力的重复性工作。构建小娜的关键技术之一是我们的搜索引擎必应。但是，小娜并不止于提供相关信息的链接，而能够通过必应来发现并提供问题的答案，以及丰富的背景信息。[10]

微软不仅使用人工智能技术来创建和增强自己的产品，也帮助其他开发者使用这些技术来构建他们自己的人工智能产品。微软人工智能平台（Microsoft AI Platform）为开发者和组织机构（不论规模大小）提供基础设施、工具与服务，帮助他们降低开发人工智能的难度。在我们提供的服务中，微软认知服务（Microsoft Cognitive Services）是一套包括视觉、语音、语言和搜索在内的预设人工智能功能，这些服务架构在云上，可以很容易地集成到应用程序中。其中一些可以定制，使用者可以基于其公司所在的特定行业和业务需求进行优化，以协助促进和改善业务流程。这些服务的覆盖范围见下图。

图4
来源：微软

我们也提供了可用于简化"机器人"创建的技术，使其可以采用更自然的方式，通过对话与人类交互。我们还在提供越来越多的编码和管理工具，以期进一步简化人工智能的开发过程。我们的基础设施服务可以帮助人们开发和部署算法、存储数据并从中获得启发。

最后，我们正在通过微软人工智能商业解决方案（AI Business Solutions）构建智能系统，帮助组织机构更好地理解其收集的信息，并采取相应行动，提高生产力。

人工智能商业解决方案的一个例子是客户关怀智能项目（Customer Care Intelligence）。目前，澳大利亚民政服务部（DHS）正在使用这一项目，并成功转变了为公民提供服务的方式。该项目的核心是一个专家系统，由一个名为"Roxy"的虚拟助理协助理赔官员回答提问和解决问题。Roxy接受了DHS运营蓝图培训，内容涵盖该机构的所有政策和程序，并输入了三个月期间内理赔官员和DHS管理者之间沟通的全部问题。在早期使用阶段，系统便能够回答近80%的提问。理赔官员的工作量预计将因此减少约20%。

由于Roxy的内部推广非常成功，澳大利亚DHS已着手开发可以直接与公民互动的虚拟助理。其中一个项目专门用于帮助高中生了解相关资格审查程序，以便其决定申请就读大学还是报名参加澳大利亚技术与继续教育学院的职业培训课程。

现代人工智能的潜力——应对社会挑战

微软的目标是通过开发人工智能系统,帮助全世界有效解决区域性挑战和全球性挑战,推动社会进步,创造更多经济机遇。

得益于人工智能,今天的人类几乎在所有领域都在以更快的速度、更大的步幅前进。作为全球经济发展的核心,数字化转型的实现至关重要。从客户沟通到转型产品,到优化运营再到赋能员工,企业或机构将从数字化转型中全方面受益。

但更重要的是,人工智能有能力帮助社会攻克最为艰巨的挑战。人类目前面临着诸多甚为复杂且亟待解决的困难:从消除贫困和改善教育,到提供医疗和消除疾病,应对可持续发展挑战(如生产足够的粮食养活快速增长的全球人口),再到提升社会的包容性。想象一下,如果能够利用人工智能来帮助我们找到解决这些挑战的办法,那么,将拯救多少生命,减轻多少痛苦,释放出多少人类潜力?

当今社会最紧迫的挑战之一,是如何以合理的成本为全球约75亿人口提供有效的医疗。无论是通过分析大量患者数据来揭示隐藏的规律,从而明确有效治疗的方向,还是发现有可能成为新药物或疫苗的化合物,或者是基于深入的基因分析，来发挥个体化医疗的潜力，人工智能提供了无限的可能，随时可能转变我们认识疾病和改善健康的方式。机器阅读能帮助医生从数千份文件中快速找到重要信息，而医生本人可能根本没有时间去阅读这些文件。这样，医务人员就可以将更多的时间用在更有价值甚至挽救生命的工作上。

安全高效的交通是人工智能可能发挥重要作用的另一个领域，也是另一个重大挑战。人工智能控制的无人驾驶汽车能够减少交通事故，扩大现有道路基础设施的容量，每年挽救数十万人的生命，还可以改善交通状况并减少碳排放。不能独立驾驶的人群，也可因此提高独立生活能力，从而促进社会更具包容性。

在教育领域，人工智能能够分析人们获取知识的方式，然后利用这些信息开发出相应的模型，来预测人们的浸进程度与理解程度。未来的教育将朝向在线教学和教师引导型教学相结合的方向发展，人工智能将引发人们学习方式上的革命性改变。

人工智能还能帮助政府改进与公民互动的方式和提供服务的方式。我们前面已经看到，澳大利亚民政服务部正是利用客户关怀智能项目的自然语言功能来回答民众的问题。

- **人工智能帮助视障人士通过倾听获取周围信息**

人工智能还可用于服务全球10亿多名残障人士,这是它可以大展拳脚的另一个领域。例如,微软最近在iOS应用程序商店中推出了一款名为"Seeing AI"的产品,帮助失明和弱视人群应对日常生活中的问题。

Seeing AI的开发团队包括一名在7岁时失明的微软工程师。这一应用程序功能强大,虽然还处于早期阶段,但已经证明了人工智能可以捕捉用户周围的图像并即时描述所发生的事情,从而提高残障人士的生活能力。比如,它可以读取标志和菜单,通过条形码识别产品、解读笔迹、清点货币、描述附近的场景和物体,或者在开会时告诉用户桌子对面的男士和女士正在微笑着认真倾听。[11]

- **人工智能帮助农民提高生产力和增加产量**

在未来25年内,全球人口预计将增加约25亿。人工智能能够提高农业产量和减少浪费,来有效地增加食品总量。例如,我们的"FarmBeats"项目即采用先进技术,将现有设施连通,通过强大的云计算和机器学习功能,来实现低成本的数据驱动型农业。该项目为农民提供了易于理解的建议,帮助其提高农业产量、降低总体成本,并减少农业对环境的影响。[12]

使用人工智能将带来巨大的效益——例如,帮助我们提高生产力和效率、完成更多工作、改善经营成果、提升政府服务效率、协助解决棘手的社会问题等——但至关重要的是，每个人都应拥有使用人工智能的机会。予力所有人和所有组织以人工智能，是让每一人抓住人工智能带来的机会，以及让人工智能惠及所有人的前提条件。

人工智能带来的挑战

人工智能建立在人类取得的伟大成就基础之上，包括电力、电话和晶体管；和这些伟大的技术进步一样，人工智能也将带来巨大的变化，甚至包括一些我们今天完全无法想象的变化。而且，正如当年的情况一样，我们也应当考虑如何解决巨变所带来的社会问题。最重要的是，我们需要共同确保以负责任的方式开发人工智能，使之赢得人们信任，得到广泛部署，从而提高企业和个人的生产力，协助解决社会问题。

我们需要就这些新技术的道德和社会影响达成共识，为制定共同的原则框架奠定基础。这些原则将指导研究人员和开发者研发新一代人工智能系统和功能，指导政府制定新的法律法规，以保护公民的安全和隐私，并确保人工智能惠及大众。

在第2章中，我们将初步探讨如何在尊重普世价值观的前提下继续前进，解决人工智能带来的各种社会问题，同时确保充分发挥人工智能的潜力，创造更多机会，改善人们的生活。

第 2 章

负责任地使用
人工智能的
原则、政策和法律

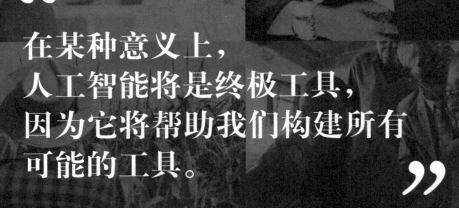

> 在某种意义上，人工智能将是终极工具，因为它将帮助我们构建所有可能的工具。

埃里克·德雷克斯勒

第2章
负责任地使用人工智能的原则、政策和法律

在 教育、医疗、交通、农业、能源和制造等领域，人工智能已初露锋芒，大大提高了人们的理解深度和决策能力，但另一方面，它也必将引发新的社会问题。例如，我们如何才能确保人工智能公平对待每一个人？如何才能最好地确保人工智能的安全、可靠？如何在保护隐私的同时，享受人工智能带来的好处？随着机器变得越来越聪明，越来越强大，我们如何才能避免失去对机器的控制？

当然，人工智能系统的打造者必须遵守世界各地的法律规定，如关于公平性、隐私权、不合理行为造成的伤害等法律。人工智能系统并非法外之地。但是，我们仍有必要制定和采取明确的原则，指导人们建设、使用和应用人工智能系统。行业组织和其他组织应在这些原则的基础上，针对人工智能系统开发的关键方面，创立具体的最佳实践，例如用于培训人工智能系统数据的性质、采用的分析技术，以及如何向使用人工智能系统的人们解释人工智能系统得出的结果。

为防患于未然，制定这些原则势在必行。否则，人工智能系统可能无法彻底取得人们的信任。如果人们不信任人工智能系统，他们也就不会为人工智能系统的开发作出贡献，更不会使用这些系统。

伦理和社会影响

商业领袖、政策制定者、研究人员、学者和非营利组织代表必须携手努力，共同确保基于人工智能的技术之设计和实施能够赢得使用者和数据收集对象的信任。微软公司作为联合创办人创立的人工智能合作组织（PAI），致力于促进关于上述问题的讨论。许多大学、政府和非营利组织也正在开展重要的工作。[13]

要设计出可信赖的人工智能，我们必须采取体现人类道德原则的解决方案，而且这些道德原则应深深根植于重大和永恒的价值观。根据我们的设想，我们将重点放在我们认为应用于指导人工智能开发的六大原则上。具体而言，人工智能系统应公平、可靠与安全、隐私与保障、包容、透明和负责。这些原则对于解决人工智能的社会影响和建立信任至关重要，因为技术作为产品和服务的一部分，已日益深入人们的日常工作和家庭生活。

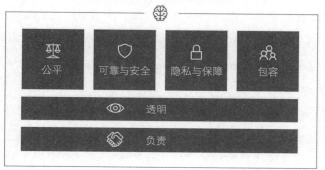

图5
来源：微软公司

公平——人工智能系统应公平对待所有人

人工智能系统应当以公正、一致的态度对待每一人，而不能对情况类似的群体施以不同的影响。例如，在提供医疗、贷款申请或就业方面的指导时，人工智能系统应为症状、财务状况或专业资格类似的人们提供相同的建议。如果设计得当，人工智能可以促进决策的公平性，因为计算机纯粹基于逻辑，理论上不会受制于有意识和无意识的偏见，但人类的决策却不可避免地会被这些偏见所牵引（影响）。然而，因为人工智能系统是由人类设计的，训练时使用的数据，反映的也是我们置身其中那并不完美的世界，因此，如果不进行严谨的规划，人工智能的运行可能失于公平。为了确保使用此项新技术的解决方案的公平性，开发人员必须了解各种偏见可能以何种方式被引入人工智能系统，以及这些偏见可能对基于人工智能提出的建议造成什么影响。

任何人工智能系统的设计均始于训练数据的选择，这是可能产生不公的第一个环节。训练数据应足以代表我们所生活的多样化的世界，至少是人工智能系统将运行的那一部分世界。试以用于面部识别或情绪检测的人工智能系统为例。如果只对成年人脸部的图像进行训练，由于面部结构的差异，这个系统就可能无法准确识别儿童的特征或表情。

但仅仅确保数据的"代表性"还不够。种族主义和性别歧视也可能悄悄混入社会数据。使用这些数据训练人工智能系统，可能会不经意地导致这些有害偏见永久存续。假设我们设计一个帮助雇主筛选求职者的人工智能系统。如果使用公共就业记录数据进行训练，该系统很可能会"学习"到大多数软件开发人员是男性。因此，在选择软件开发人员职位的人选时，该系统就很可能偏向于男性，尽管实施该系统的公司想要通过招聘提高员工的多样性。[14]

如果人们不了解系统的局限性，特别是，如果人们假定技术系统比人更少出错、更加精准，因此更具权威，那么人工智能系统也可能造成不公。在许多情况下，人工智能系统的输出实际上是一个预测。例如，"申请人贷款违约概率为70%"。人工智能系统给出的结果可能非常准确，即，如果银行每次都向"违约风险率"为70%的人提供信贷，那么实际上，这些人中将有70%的人违约。但是，如果贷款管理人员错误地将"70%违约风险"简单地解释为"不良信用风险"，并拒绝向所有这些人员提供信贷,那么,其中有近三分之一的申请人,虽然信用状况良好,其贷款申请仍将被拒绝,这样就会导致不公。因此,我们同样需要对人进行培训,使其理解人工智能结果的含义和影响,以便运用人类的良好判断力,弥补人工智能决策中的不足,这一点至关重要。

如何才能确保人工智能系统公平对待每个人?几乎可以肯定的是,在该领域内,我们每个人都需要进行大量学习,而且必须持续开展研究,促进积极讨论,共享新的最佳实践。但是,在目前,一些重大的主题才刚刚开始浮现。

首先,我们认为,人工智能系统的设计者应反映我们所生活世界的多样性。我们还认为,对相关主题具有专业知识的人员(例如,那些具有消费者信用专业知识的人员对应人工智能信用评分系统)应参与人工智能的设计过程和部署决策。

其次,如果人工智能系统的建议或预测被用于帮助作出与人相关的决定时,这些决定必须由人类主导作出,这一点至关重要。同样有必要的是对相关研究进行投入,以更好地了解人工智能系统对人类决策的一般性影响。

最后——也是至关重要的一点——业界和学术界应继续推进目前正在开展的前景广阔的工作,开发分析技术,以检测和解决潜在的不公平问题,例如人工智能系统训练数据进行系统性评估的方法,以确保这些数据具备适当的代表性,并书面记录与数据来源和特征相关的信息。

从根本上说,要解决人工智能系统中可能存在的偏见,全面界定所需开展的各项工作,需要广泛纳入各个利益相关方,持续开展讨论。学术研究界作出的努力,包括研究人员年度大会上所重点讨论的机器学习的公平、问责和透明度等问题,提升了公众对这一问题的意识。我们鼓励公共、私营和民间部门加大工作力度,扩大讨论范围,帮助找到解决方案。

可靠——人工智能系统应确保运行可靠、安全

人工智能技术的复杂性,引发了人们对人工智能的担心,担心人工智能系统可能会在不可预见的情况下造成伤害,或者被人恶意操纵,实施有害行为。与其他任何技术一样,人工智能能否取得人们信任,最终取决于基于人工智能的系统能否可靠、安全和一致地运行——不仅是在正常情况下,更包括在不可预见的或受到攻击的情况下。

所以,必须有明确的设计原则,即人工智能系统在可预见的情况,必须在一整套行为规范内运行。由于人工智能系统由数据驱动,其行为方式以及其是否有能力在复杂状况下可靠而安全地运行,很大程度上取决于开发人员在设计和测试过程中所预期到的情况和环境。例如,用于检测错位物体

的人工智能系统，在照明不足的条件下，很可能难以识别物体，因此，设计人员应该在普通光线和暗光环境中都进行测试。为确保系统能够安全地对意外的情况作出反应，同时不造成意外性能故障，更不会以与预期不符的方式进行演变，在系统开发和实施过程中，严格的测试是必不可少的。

在设计和测试过程中，还应预料到并防止意外系统交互，或不法行为者影响运行（如网络攻击）的可能性。要保护人工智能系统的安全性，开发人员需要识别异常行为，并防止被操纵，例如导入可能对人工智能行为产生负面影响的恶意数据。

此外，人工智能的目的在于提高和增强人类的能力，因此，人工智能系统的部署方式、部署时间，以及是否适宜继续使用人工智能，这些关键决定均应由人类主导。人工智能系统往往并不能看到或理解更大的社会图景，因此，人类的判断对于识别人工智能系统潜在的盲点和偏见非常关键。开发人员在构建和布置系统时，应意识到这些挑战，与客户共享相关信息，帮助客户监控和理解系统行为，以便客户快速识别和纠正任何可能出现的意外行为。

例如，在人工智能研究领域，一个旨在帮助决定肺炎患者是否需要住院的系统"学习"到哮喘患者死于肺炎的概率低于一般人群。这一结果令人惊讶，因为人们通常认为，哮喘患者比其他人更容易死于肺炎。系统发现的这一关联性并没有错,但是它并未注意到,哮喘患者死于肺炎的死亡率较低的主要原因是他们的风险更高,因此能够比其他患者更快和更全面地获得护理。如果研究人员没有注意到人工智能系统作出了一个误导性的推论,那么该系统就可能会建议哮喘患者不要进行住院治疗,而这一结论与数据揭示的真相背道而驰。[15]从这个例子,我们可以明显地看到,在人工智能系统的开发和部署过程中,有必要由人类(特别是具有专业知识的人员)进行观察和评估。

在其他工程领域,已经确立了稳健和故障安全的设计原则;对于设计和开发可靠和安全的人工智能系统,这些原则同样具有极高的指导意义。随着人工智能系统在交通、医疗保健和金融服务等领域的应用日趋广泛,要进一步提高人工智能系统的安全性和可靠性,有必要联合行业参与者、政府、学者和其他专家开展研究、协作,而且这一工作将越来越重要。

我们相信,以下措施可以提高人工智能系统的安全性和可靠性:

▸ 对用于训练和运行人工智能产品和服务的数据和模型,进行系统性的质量和适当性评估,并系统地分享关于训练数据潜在不足的信息。

- 通过相关流程对人工智能系统的运行情况进行记录和审计,以帮助理解持续进行的性能监控。
- 当人工智能系统用于作出与人相关的相应决定时,应要求对整体系统运行提供充分说明,特别是有关训练数据和算法、已发生的训练失败以及产生的推论和重大预测等。
- 人工智能被用于作出与人相关的决定的,应让相关领域的专家参与设计和运行。
- 评估在危急情况下,人工智能系统应何时、以何种方式寻求人类帮助,以及人工智能控制的系统如何以有意义和简明的方式将控制权移交给人类。
- 建立强大的反馈机制,以便用户轻松报告遇到的性能问题。

创建安全、可靠的人工智能系统是我们共同的责任。因此,行业参与者应当共享设计和开发的最佳实践,例如有效测试,以及试验和报告的架构。人机交互以及人工智能驱动系统在发生故障时如何将控制权移交给人类等话题,不仅是当前研究的重要领域,对于加强行业内的协作和交流也具有重要意义。

隐私与保障——人工智能系统应当有保障且尊重隐私

随着我们的信息越来越广泛地以数字化形式存储，如何保护我们的隐私和个人数据，正在变得越来越重要，也越来越复杂。固然，隐私和安全保障对于所有技术开发都很重要，但人工智能近期的发展，促使我们更加关注这些问题。这是赢得公众信任的必要条件，公众信任又是充分发挥人工智能作用的前提。简而言之，除非人们确信自己的隐私能够得到保护，并且其数据有安全保障，否则没有人会共享有关自己的数据（而这些数据是人工智能帮助人们作出决策的必要条件）。

隐私不仅是商业上的需要，也是所有云计算服务取得信任的关键基础。因此，微软坚定地承诺保护客户数据的安全和隐私，我们也正在升级工程系统，确保对全球数据保护法律的合规，包括欧盟的《通用数据保护条例》（GDPR）。微软正在投资建设基础设施和系统，以使我们能够在有史以来规模最大的开发工作中符合监管环境要求，实现对GDPR的合规。

与其他云技术一样，人工智能系统必须遵守隐私法律；这些法律针对数据收集、使用和存储提出了透明度要求，并要求向消费者提供适当的控制权，以便其选择数据的使用方式。在设计人工智能系统时，也应确保隐私信息的使用符合隐私标准，并防止不法行为者窃取隐私信息或造成损害。应针对以下各项开发和实施相应行业流程：跟踪客户数据的相关信息（例

如数据收集时间和适用条款）；数据的访问和使用；访问和使用的审核。微软仍在投入开发、完善合规技术和流程，确保以负责任的方式处理人工智能系统收集和使用的数据。

我们需要的是一种能够促进技术开发和政策制定的做法，在保护隐私的同时，促进对必要数据的获取，让人工智能系统能够获取必要的数据，从而有效运行。在不断创造并推进隐私保护的尖端技术的研究和发展方面，微软始终是全球的引领者。这些技术包括差分隐私（differential privacy）、同态加密、区分数据与个人身份识别信息的技术，以及防止滥用、黑客攻击或篡改的技术等。我们相信，这些技术可以降低人工智能系统侵犯隐私的风险，使得人工智能系统可以在不访问或不知道个人身份的情况下使用个人数据。微软将继续对研究进行投入，与政府和业界其他机构合作，开发可以根据敏感度及拟定用途来进行部署的、切实有效和高效率的隐私保护技术。

包容——人工智能系统应确保人人赋能、人人参与

要确保人工智能技术造福、赋能每一个人，这些技术必须考虑广泛的人类需求和经验。包容性设计将帮助系统开发人员了解和解决产品或环境中存在哪些潜在障碍，可能导致无意中将部分人群排除在外。因此，所设计出来的人工智能系统应当能够理解其使用者的情境、需求和期望。

信息和通信技术对世界十亿残疾人所具有的重要意义毋庸多言。已有超过160个国家批准了《联合国残疾人权利公约》，其中就涵盖了教育和就业领域数字技术的可及性。

在美国，《美国残疾人法》和《通信与视频无障碍法》对技术解决方案设定了无障碍访问要求，联邦和州法规也强制要求采购无障碍技术。欧盟法律也有相关规定。人工智能作为一项强大的工具，可以有效增强信息、教育、就业、政府服务，以及社会和经济机会的可及性。例如，实时语音文本转录、视觉识别服务，以及为人们提供输入建议的预测文本功能，都是人工智能为听视障和其他残疾人士服务的例证。

我们还相信，人工智能如果同时具备情感智能和认知智能，其积极影响将得到最大程度的发挥；这种平衡可以提高其可预测性和理解力。例如，基于人工智能的个人助理可以通过确认并且在必要时自我纠正对用户意图的理解，同时识别和调整对用户而言最重要的人员、地点和事件，来展现对用户的认知。个人助理应根据具体情境和用户预期的方式提供信息和建议，还应提供相关信息，帮助人们理解系统对他们所作出的推断。随着时间推移，这种有效的互动将会增加人工智能系统的使用频率，并逐渐赢得人们对其性能的信任。

透明——人工智能系统应易于理解

以上四项价值观的基础是两项基本原则：透明和负责。这两项基本原则是其他所有原则的基础。

一旦人工智能系统被用于作出影响人们生活的决策时，人们就有必要了解人工智能是如何作出这些决策的。一种方法是提供解释说明，包括提供人工智能系统如何运行以及如何与数据进行交互的背景信息，这种方法可以帮助与用户以及受这些系统影响的人员建立信任。通过这些信息，人们将会更容易识别和意识到潜在的偏见、错误和意想不到的结果。

仅仅发布人工智能系统的算法很难实现有意义的透明度。最新（通常是最有发展前途的）人工智能技术，例如深度神经网络，通常没有任何算法输出可以帮助人们了解系统所发现的细微模式。有鉴于此，我们需要一个更全面的方法，使人工智能系统设计人员能够尽可能完整、清晰地描述系统的关键组成要件。

微软正在与人工智能合作组织（PAI）及其他组织合作开发最佳实践规范，以实现人工智能系统切实的透明度。包括通过上文所述的实践规范以及各种其他更易于理解的方法、算法或模型，来替代那些过于复杂且难以解释的方法。要理解机器学习模型如何工作，开发出新技术，来提供更有意义的透明度，需要对这一领域开展进一步研究。

负责

最后,与其他技术和产品一样,设计和部署人工智能系统的人员必须对其运行的系统负责。为了建立人工智能的问责标准,我们应借鉴包括医疗和隐私在内的其他领域的经验和规范。人工智能系统的开发者和用户应考虑这些规范,并定期检查他们是否遵循了这些规范,以及这些规范是否产生了应有的效果。对于应采取哪些规范来帮助解决上文讨论的问题,以及与人工智能系统开发和布署相关的特别重大问题,可以由内部审查委员会提供监督和指导。

内部监督与指导——微软人工智能工程与研究道德委员会（AETHER）

最后,为确保上述六项原则有效实施,需要将这些原则与日常运行紧密结合。在微软,我们解决这一问题的途径之一是设立开发和研究人工智能与道德标准（AETHER）委员会。作为一个新成立的内部组织,该委员会囊括了来自微软公司开发、研究、咨询和法律部门的高管,专注于积极制定内部政策,以及如何应对所出现的具体问题。AETHER委员会审查和界定最佳实践,提供指导原则,指导微软人工智能产品和解决方案的开发和部署,并帮助解决微软人工智能研究、产品和客户沟通中产生的伦理和社会影响问题。

制定人工智能相关的政策和法律

几乎在人类涉足的所有领域，人工智能都可以作为催化剂，加快人类前进的步伐。但是，任何使我们超越现有知识和经验的创新，都会引出一些重要议题，如人与技术之间的关系，以及新技术驱动的能力对个人和社区的影响等，人工智能也不例外。

在我们生活的世界中，人工智能扮演的角色将日益重要，而我们是面对人工智能的第一代人类。可以肯定的是，大多数现行标准、法律和法规并非专门针对人工智能制定。但是，尽管现有规则不是专门为人工智能而制定，但这并不意味着基于人工智能的产品和服务不受管制。例如，保护个人信息隐私和安全、管理数据流和使用方式、促进消费者信息使用的公平性或关于信用或就业决策的现行法律，大致适用于数字产品和服务或其在决策中的应用，无论这些法律是否明确提到了人工智能功能。例如，基于人工智能的服务不能豁免与GDPR有关的要求，或者美国保护医疗保健数据隐私的《健康保险流通与责任法案》或现行的汽车安全法规。

随着人工智能的继续发展，决策者不仅要监督其影响，还需要解决新的问题，并更新相关立法。其中一个目标应是确保政府与企业和其他利益相关者合作，确保在最大限度地发挥人工智能的潜力与改善人民生活的同时，有效应对新出现的挑战，并在两者之间求取适当的平衡。

正因为如此，"人工智能法"似乎不可避免地成为了一个重大和全新的法律课题。但是，具体的时间框架如何？该领域又应当以何种方式发展和演变？

我们认为，为所有利益相关方提供充分的时间，以识别和阐明关键原则，指导负责任并可信赖的人工智能的开发，并通过采用和改进最佳实践实施这些原则，可以实现最有效的监管。在制定新的法规或法律之前，需要首先在一定程度上厘清必须解决的基本问题和原则。

美国和欧洲信息隐私法律的演变提供了一个可以借鉴的模板。1973年，美国卫生、教育及福利部（HEW）发布了一份综合报告，因计算机引发的信息化以及联邦机构持有的个人数据库日益增长所引发的一系列社会问题。[16]该报告提出了一系列重要原则，即《公平信息处理条例》，试图描绘基本的隐私保护理想架构，而不考虑涉及的具体情境或技术。在报告发布后的几十年中，这些原则（主要归功于其根本性和普遍性）帮助制定了一系列联邦和州法律，对教育、医疗、金融服务和其他领域对个人信息的收集和使用进行监管。在这些原则指导下，美国联邦贸易委员会（FTC）制定了一套隐私案例法，以防范影响商业的不公平或欺骗行为。

在国际上，《公平信息处理条例》影响了欧洲多个国家和地区法律的发展，包括德国和法国在内；在许多方面，德国和法国已成为隐私法律发展的领军者。从20世纪70年代末开始，经济合作与发展组织（OECD）以《公平信息处理条例》为基础，颁布了具有划时代意义的《隐私保护指引》（Privacy Guidelines）。与美国卫生、教育及福利部的《公平信息处理条例》一样，经合组织创建的示范规定所具有的普通适用性和可扩展性，直接引领了欧盟1995年的全面《数据保护指令》及其后续的《一般数据保护条例》。

美国和欧洲的法律最终走向了不同的道路，美国趋向于采取部门法的方式，而欧盟则采取了更加综合的监管方式。但是，上述两者均建立在普适的基础性概念之上，也部分立足于现有法律和法律原则。这些规则涉及非常广泛的新技术、用途和商业模式，也处理了日益多样化的社会需求和期望。

时至今日，我们认为，政策讨论的重点应放在基本人工智能技术的持续创新和发展上，支持人工智能在不同行业的开发和应用，基于以人为本的人工智能的共同愿景，鼓励与之相符的结果，促进最佳实践的发展和共享，增强人工智能的可信任性和负责性。在决策者为实现这些目标而制定相应框架制度时，我们建议考量如下因素：

数据的重要性

从近期来看，大部分人工智能政策和监管议题的重点似乎在于数据的收集和使用。要开发更为有效的人工智能服务，数据不仅必不可少，而且相关数据更是多多益善。数据的访问和使用涉及广泛的政策问题，从确保个人隐私保护、保护敏感和专有信息，到一系列新的竞争法问题等。要对这些目标进行慎重和卓有成效的平衡，需要政府、行业参与者、学术研究人员和公民社会之间进行广泛的讨论与合作。

一方面，我们认为，政府应倡导通过约定俗成的方法，为机器学习提供广泛的数据，从而加速人工智能的发展。大量有用的数据来自公共数据集，这些数据属于公众本身。各国政府还可以投资和推广各项方法和程序，将公共和私营组织的相关数据集连接和组合在一起，同时根据具体情况的要求，确保数据的保密、隐私和安全。

与此同时，各国政府必须在考虑数据类型以及数据使用情境的基础上，制定和推广有效的隐私保护方法。为了帮助降低侵犯隐私的风险，政府应支持和促进技术发展，使得系统能够在无需访问或了解个人的身份的情况下使用个人数据。开展进一步研究，加强"去身份识别"技术，以及持续开展关于如何平衡身份再识别风险与社会效益的讨论，均具有重要意义。

决策者希望更新数据保护法时，应对数据可以提供的好处与重大的隐私权益进行慎重权衡。虽然通常而言，一些敏感个人信息（例如社会保障号码）应受到高度保护，但也应避免采用过于严苛的方法，因为个人信息的敏感度往往取决于提供和使用信息的情境。例如，公司名录中的员工人名通常不被认为属于敏感信息，因此所需的隐私保护力度很可能低于收养记录中出现的人名。一般而言，在对法律进行更新时，立法人员应认识到，要保护公共利益（例如防止传染病蔓延及其他严重的健康威胁），敏感信息的处理可能会越来越关键。

竞争法是另一个相关的重大政策领域。智能设备、应用程序和云服务产生的数据规模如此之大，以致人们越来越担心信息会日益集中在少数几家公司手里。但是，除了公司从客户获取的数据以外，另外还有公开可用的数据。政府可以确保人工智能开发者以非排他性的方式使用公共数据，帮助增加可用数据的数量。这些措施将有助于各类开发人员更好地利用人工智能技术。

与此同时，各国政府应监测独特数据集（即无法替代的数据）的访问权限是否正在成为竞争的障碍，以及是否需要解决这一问题。其他问题包括数据是否过度集中在少数公司手里，以及竞争对手能否借助复杂的算法有效"确定"价格。所有这些问题都值得关注，但都可能在现行竞争法的框架内得到解决。在一家公司试图收购另一家公司时，会直接

第2章
负责任地使用人工智能的原则、政策和法律

出现数据可用性问题,竞争主管部门需要考虑合并之后的公司是否会拥有极具价值的独特数据集,导致没有其他公司能够有效开展竞争。当然,这种情况出现的可能性并不太高,因为数字技术产生的数据规模庞大,多家公司拥有相同数据的情形非常常见,而且人们通常会使用多种服务,从而为许多不同的公司生成相应数据。

算法可以帮助提高价格透明度,这有助于企业和消费者以最低的成本购买产品。但是,很可能有一天算法会变得非常复杂,使用这些算法的公司或许可以在没有进行约定的情况下,确定完全相同的价格。竞争主管部门需要慎重研究价格透明度的好处,以及随着时间推移,透明度可能降低恶性价格竞争的风险。

促进负责任和有效地使用人工智能

除了解决数据相关问题之外,政府还扮演着另外一个重要角色,即如何促进以负责任和有效的方式利用人工智能。对于这一问题,应从在公共部门部署负责任的人工智能技术入手。这样,既可以更有效地为公民提供服务,同时,也能为政府提供第一手经验,帮助其制定最佳实践,以遵循上文所述的道德原则。

此外,各国政府的另一个重要责任在于资助核心研究,进一步推动人工智能发展，支持跨学科研究，重点是针对人工智能技术实施后可能出现的社会经济问题，研究和制定相应的解决方案。这种跨学科研究对于未来人工智能法律法规的设计也具有宝贵价值。

各国政府还应激励在广泛的行业和各种规模的企业中采用人工智能技术，特别是为中小企业提供激励机制。人工智能为小企业提供的助力，可以促进经济增长和创造发展机遇，并在解决收入停滞，以及缓解随着收入不平等加剧而出现的政治和社会问题方面发挥重要作用。政府采取这些措施后，可以进一步采取保障措施，确保人工智能不会被有意或无意地以相关法律禁止的方式用于歧视。

责任

政府还必须在支持创新的同时，保证消费者安全，并在两者之间求取平衡。为此，人工智能系统制造者因不合理做法而造成损害的，政府应向其究责。过失法原则已经发展得极为成熟，人工智能系统的实施和使用所引起的损害，最适合使用这一原则进行处理。这是因为这些法律原则鼓励合理的行为，不要求不符合责任认定标准的行为人承担责任。因此，这特别适合于人工智能领域，原因如下：首先，人工智能系统可以发挥的潜在作用及其可以带来的好处十分巨大；其次，各种自动化系统以及许多其他现有和预期的人工智能

技术和服务，已经为社会所熟悉；再次，目前已开展了大量工作，以协助降低这些系统造成损害的风险。

在软件领域，过失标准已得到普遍采用；通过这一标准来认定人工智能造成的损害责任方，是政策制定者和监管者平衡创新与消费者安全的最佳方式，并且可以提高技术开发者和用户所感知的确定性。这也有助于企业对其自己的行为负责，协调为企业提供的激励措施以及承担的损害赔偿责任。

促进对话和分享最佳实践

为了最大限度地发挥人工智能的优势，同时减少风险，尽量避免意外后果，我们必须继续在政府、企业、非盈利组织和公民社会代表、学术研究人员和所有其他利害相关的个人和组织之间举行公开讨论。通过各方共同努力，我们可以识别具有明显社会或经济后果的问题，优先制定保护人类的解决方案，同时避免不必要地限制未来的创新。

要解决当前和未来问题，一项有益措施是开发和分享创新的最佳实践，以指导以人为本的人工智能的创造和实施。行业领导组织，例如人工智能合作组织，将行业和非营利组织联合起来后，作为制定和发布最佳实践的论坛。通过鼓励开诚布公的讨论，协助分享最佳实践，政府还可以帮助在人工智能开发者、用户和广大公众之间创造一种合作、信任和开放的文化。此项工作可作为未来制定的法律和法规的基础。

此外，非常关键的一点是，我们应当承认，人工智能技术对工作和工作性质的影响，已经引发了广泛关切，我们应该采取措施确保人们为迎接人工智能对工作场所和劳动力产生的影响做好准备。人工智能已经在改变企业与员工之间的关系，改变人们的工作方式、时间和地点。随着变化的速度越来越快，新技能的重要性将日益凸显，而且必须通过全新的途径，来为人们提供培训和就业机会。

在第3章中，我们将讨论人工智能对职业和工作产生的影响，并建议采取相应措施，为处于不同年龄段、不同求学阶段和职业阶段的人们提供教育和培训，帮助人们把握人工智能时代的机遇。我们还将探讨在员工和雇主之间的关系迅速变化的时代，对员工权益保护和社会保障进行反思的必要性。

第 3 章

人工智能与职业和工作的未来

第3章
人工智能与职业和工作的未来

250多年以来，技术创新一直在不断改变职业和工作的性质。18世纪60年代，第一次工业革命爆发后，工作集中迅速地从原来的家庭和农场转移到快速发展的城市。19世纪70年代兴起的第二次工业革命延续了这一趋势，并诞生了生产线、公司，形成了当今办公室的雏形。交通工具由马车向汽车转变，既淘汰了许多职业，也创造了人们之前从未想象的新工种。[17]在势不可挡的经济变革席卷之下，部分工作条件变得十分艰苦甚至极度危险，为此，政府不得不采取相关劳动保护措施和实践，并一直沿用至今。

第三次工业革命的兴起在数十年之前，我们大部分人对其带来的变化都有切身体会。就微软而言，我们实现了公司最初的愿景——让每张办公桌和每个家庭都拥有一台计算机。这一变革将信息技术引入工作场所，改变了人们工作的沟通和合作方式，创造了新的IT岗位，但也在很大程度上让许多秘书失去了工作，因为手动输入手写文件的需求已不复存在。

现在，信息技术再度发生了变革，职业和工作的性质也再次随之改变。虽然目前可用的经济数据还远不完善，但已经可以清楚地看到，企业组织工作、人们求职的方式和工作需要的技能正在发生显著变化。未来几十年，变化的速度可能会越来越快。

这一变化的主要推动力来自于人工智能和云计算。从快速发展的"按需"经济或"零工"经济中，可以很明显地看到这点。在"按需"经济或"零工"经济模式下，数字平台对劳动者的技能与消费者或企业的需求进行双向匹配，为劳动者提供了几乎能够在全球各地工作的机会。人工智能和自动化已经开始决定未来哪些职业或工作内容能够继续存在。有人估测，未来10年，将有多达510万份工作消失；不过，人工智能和自动化也将创造新的经济机会，以及全新的职业和工种。[18]

上述工作性质的根本性变化，要求我们采用全新的思维方式来看待技能和培训，确保劳动者从容应对未来的变化，并确保关键工种拥有充分数量的人才储备。同时，我们也需要完善教育生态系统，鼓励员工采取终生学习的态度，使每个个体掌握只有人类才能掌握的技能，并将持续教育融入全职工作和按需工作中。对企业而言，要重新思考招募和评估人才的方法，扩大求职者的范围，并采用各种工作组合评估其能力和技术。用人单位还要更加注重为其现有劳动力提供在职培训、学习新技能和获得外部教育的机会。

我们不仅要重新思考劳动者的培训方式，确保劳动者跟上时代要求，还要考虑传统雇佣模式巨变对劳动者产生的影响，因为传统雇佣模式下，用人单位通常会向劳动者提供福利和保护措施。工作方式快速转变，可能削减劳工保护措施和福利，包括失业保险、劳工赔偿以及社会保障制度（就美国而言）。为了避免这一情况，我们需要改革就业相关法律制度，认识到所出现的全新工作方式，提供充分的劳工保护，并维护社会保障制度。

技术对职业和工作的影响

从古至今，每种新技术的产生，无不伴随着严峻的失业信号。例如，1928年，《纽约时报》曾刊登了一篇标题为"机器发展将造就大量闲散人口"的文章。[19]但事实是，相较被淘汰的职业，新技术创造的职业更多。以蒸汽机为例，蒸汽机的发明促进了蒸汽机车的产生，进而推动了以农村人口为主的农业社会，过渡到以制造业和交通运输业为主的城市型社会。这一转变也改变了人们工作的方式、时间和地点。以离我们较近的发明——自动取款机——为例。自动取款机替代了银行柜员的许多工作。从1988年到2004年，美国每个分行柜员的平均数量由20人降至13人。[20]尽管如此，由于柜员人数减少，分行的运营成本随之降低，银行因此有能力设立更多分行，导致了员工总数增加。自动取款机没有淘汰原有的柜员职业，而是消除了常规性工作，使柜员能够专注销售和客户服务。[21]

这一模式几乎可以在所有行业得到印证。最近，某经济学家对劳动力市场的分析发现，1982年至2002年间，使用计算机的职业，就业增长速度明显快于不使用计算机的行业。这是因为采用自动化技术之后，劳动者能够将精力集中在不需要重复劳动的工作内容之上；另一方面，这也增加了对执行价值更高、但尚未实现自动化任务的人类劳动者的需求。[22]

最近，公众争论的主要焦点在于自动化和人工智能对就业的影响。尽管"自动化"和"人工智能"经常被当作同义词使用，但严格来说，这是两种完全不同的技术。自动化技术通过设定程序，使系统完成特定的重复性任务。例如，原来由人工在打印机上完成的工作，通过文字处理系统实现了自动化；条形码扫描仪和POS系统，则将原来由零售业员工完成的任务进行了自动化。而人工智能的设计目的则是寻找一种模式，总结经验并作出适当的决策——无需设计明确的程序化路径，指示其如何对面临的情况作出反应。在自动化和人工智能的合力下，工作性质正以前所未有的速度发生改变。正如一位评论员所言，"自动化机器整理数据，人工智能系统则'理解'数据，这是两种截然不同、但又完美互补的系统。"[23]

由于人工智能能够补充并加速自动化,各国的决策者都意识到,在未来几十年,人工智能将成为驱动经济增长的一个重要因素。例如,中国最近就宣布,计划成为人工智能领域的全球领导者,以增强经济实力,创造竞争优势。[24]

依赖数据和信息的企业或机构——在当今时代,这就意味着几乎所有企业和机构——都能受益于人工智能。人工智能系统不仅能提高效率和生产力,还能创造具有更高价值的服务,驱动经济增长。但是,自第一次工业革命开始,所有新技术的引入,无不引发了人们的忧虑——这一技术将对职业和就业产生什么影响?在这点上,人工智能和自动化并非特例。从表面上看,人工智能和自动化确实引发了非常严峻的问题,可能在发达国家造成严重失业。微软最近委托的一项调查表明,参与调查的所有16个国家中,无一例外将人工智能对就业的影响视为一项关键风险。[25]随着机器逐渐胜任复杂分析和具备自行判断能力,人工智能是否会加剧自动化引发的失业情况,已引发广泛忧虑。

虽然我们尚不能断言人工智能是否会比以前的技术进步更具破坏力,但人工智能对职业和就业的影响是毋庸置疑的。就像之前的重大技术变革时期一样,目前我们也很难预测究竟有多少职业将受其波及。牛津大学一份被广泛引用的研究报告声称,受计算机化影响,美国有47%的岗位面临失业风险。[26]世界银行在一项研究中预测,经济合作发展组织成员国57%的职业可能实现自动化。[27]在最近一篇关于机器人与职业的论文中，研究人员指出，每使用一台机器人，就会造成就业率下降6.2‰，工资下调0.7个百分点。[28]

许多行业均面临着人工智能和自动化的双重冲击。例如，一家总部位于旧金山的公司开发了一个名为"Tally"的应用程序，能够自动核查杂货店的货架，确保商品适时上架并附有定价[29]；目前，亚马逊运营中心使用的机器人超过10万台，并正在推出无人便利店；澳大利亚的一家公司发明的机器人每小时可以砌1000块砖（而人工需要一天或更长时间才能完成）；呼叫中心使用聊天机器人来回答客户支持问题；甚至在新闻领域，体育赛事摘要的编辑工作目前也已自动化。[30]

有些职业虽然未被人工智能完全取代，但受到的影响也不可小觑：在仓储行业，雇员的工作已经从码放货物转变为操控机器人；在法律行业，律师助理和法官助理开始使用"e-discovery"软件来检索文件；在医疗行业，机器学习能够帮助医生加快疾病诊断速度，老师们也可借此更有效地评估学生的学习情况。人工智能正在重塑这些职业，但这些职业并没有消失——最根本的原因是这些职业在某些方面无法自动化。由于人工智能和机器不具备人类的创造力、合作、抽象和系统思维、复杂沟通以及在多样性的环境中工作等能力，因此，许多职业仍无法离开需要具备独特人类技能的劳动者。

一方面，人工智能会消灭、改变某些职业，另一方面，它也将创造出全新的职业。研究机构Forrester在其最近的一份报告中指出，到2027年，人工智能将替代2470万份工作，并创造出1490万份新工作。[31]由于人工智能改变了人们工作的方式，以及人们对周围世界的需求，新的职业将应运而生，而且这些职业大多集中在技术领域。例如，今后银行将不再需要柜员，但网络工程师必不可少；零售商不需要实体店的迎宾员或销售员，但需要具备Web编程技能的人员打造线上购物体验；同样，农场不再需要水果采摘员，但农业数据分析师却是必备。未来，对数据科学家、机器人专家和人工智能工程师的需求将大幅增加。

此外，人工智能也将创造出一些我们目前毫无概念的职业。尽管我们很容易识别出在哪些领域，自动化将减少对劳动者的需求，但却无法预测未来可能发生的所有变化。正如某位经济历史学家所言，"如同我们之前无法预测现在出现的职业，我们现在也无法预测未来会出现的职业"[32]。

人工智能和自动化引发的剧烈震荡，后果之一是许多行业关键人才缺失。由于各种工作对技术性技能的需求都越来越强，企业纷纷展开竞争，争夺具备专业技能，能够支持机器人学、增强现实计算、网络安全和数字科学等数字能力的员工。据估测，到2020年，受人才短缺影响，30%的技术类职业将出现岗位空缺[33]，如果把引进技能培训方案所需的时间考

第3章
人工智能与职业和工作的未来

虑在内，这一缺口很可能继续扩大。据世界经济论坛统计，许多学术领域的核心课程发生了前所未有的变化；对技术类专业学生而言，他们在大学第一年学到的学科知识，约50%在毕业前将会过时；到2020年，大多数职业要求具备的技能，超过三分之一可能是我们目前认为并不重要的技能。[34]广而言之，未来，技术发展将对所有工种的技能要求产生重大影响。为了顺利应对这些趋势，我们需要确保劳动力市场的参与者不断学习并获得新技能。

研究人才短缺现象和有被自动化取代之虞的、所谓"中级技能"职业的经济学家，对人工智能等技术进步表示担忧——这些技术进步正逐渐扩大具备技术性技能、接受技术培训的人群与不具备技术性技能、未接受技术培训的人群之间的收入差距。[35]随着数据分析等领域的专门技能在许多职业中发挥的作用越来越关键，机器能够自动化处理的重复性工作也越来越多，未来，对具备高度技能的劳动者的需求将增加，对不具备相应技能的劳动者的需求则将减少——这一现象被称为"技能歧视性的技术变革"。例如，在美国，从1989年到2016年，对于获得本科学历的人群而言，工作岗位增加了一倍；对于持有高中或以下文凭的人群，工作岗位则减少了13%。而在同一时期内，获得本科学历的人数增加不到50%，相较具有本科学历的人群，未获得本科学历的人群的失业率骤增了300%。[36]为了缩小这一不断扩大的差距，我们需要转变对教育和培训的思考方式，让更多的劳动者具备把握机遇的能力。

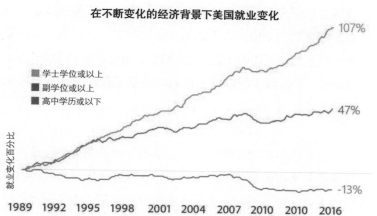

图6
来源：乔治城大学教育与劳动力研究中心

工作、工作场所和职业的性质不断变化

以前，在传统的雇佣关系模式下，人们多在办公室、工厂、学校、医院、其他商业设施等特定的工作地点工作。现在，由于远程工作和兼职工作人数激增（包括外包或项目制劳动关系等），这一传统模式正被颠覆。

部分研究表明，在2005年至2015年之间，处于弹性工作关系中的人数——包括独立承揽人和按需劳动者——从10%增加到16%，几乎构成了同一期间内净职业增长的总数。[37]麦肯锡全球研究院最近发布的一份研究表明，"采取独立工作方式的劳动力规模，高于此前认定的数字"；其中，欧洲和美国从事独立工作的人数高达1.62亿，占劳动年龄人口的20%或30%。半数以上的独立工作者以独立工作作为其主要收入来源的补充。

技术进步推动了弹性工作安排的产生。在这一领域，最显著的趋势可能是按需经济的兴起。就其本质而言，按需经济指人们通过线上人才平台或求职中介求职，为多个客户提供服务的工作安排。麦肯锡全球研究院的研究表明，15%的独立工作者使用数字人才平台联系工作。据牛津大学马丁学院技术与就业项目的研究人员估计，在未来20年内，美国约30%的职业可纳入任务制工作范畴。[38]

按需经济为劳动者和企业提供了大量机会。麦肯锡全球研究院估测，到2025年，为劳动者提供工作机会匹配服务的数字平台，对全球GDP的贡献可能高达2%，为全球创造的就业机会等同于7200万份全职工作。以下是按需经济可能带来的部分好处：

▶ 劳动者可以通过数字平台获得工作机会，而不必被迫在物理上迁往有工作机会的地点。按需工作既可以为所居住地区工作机会有限的劳动者提供帮助，又能使企业获得更大的人才库。

- 根据汉密尔顿项目（Hamilton Project）的调查，超过70%的非劳动力市场参与者表示，看护需求、身体残疾或提前退休等因素使他们无法加入劳动力大军。按需工作具有弹性特点，可减少传统雇佣模式的固有障碍。[39] 皮尤研究中心（Pew Research Center）的一项调查显示，约50%的按需劳动者表示"他们需要掌控个人的时间安排"，25%表示"其居住地没有其他工作"。[40]

- 按需经济增加了兼职劳动者的工作机会。相较全职工作，现在许多劳动者更喜欢工作方式灵活的兼职工作。[41] 对千禧一代而言，弹性工作、工作/生活平衡及其工作产生的社会影响，比高薪或全职工作更加重要。许多在婴儿潮时期出生的劳动者选择推迟参加工作的时间，且选择兼职工作的比例很高。

- 按需经济使企业能够聘用短期劳动者，有助于提升企业的灵活性，降低长期维护员工的成本。对于没有经济能力雇佣大量全职劳动力，而可以通过针对性的按需聘用安排完成工作的小型企业来说，按需经济尤其有利。由于线上平台能够为各个项目提供有竞争力的竞价，通过线上平台招募兼职人员可进一步降低企业成本。

- 按需经济使企业有机会获得内部员工缺少的技能。企业通过聘用兼职人员，能够发现具备特定技能的个体，并根据需要聘用。

- 按需经济可为劳动者带来额外收入。例如,线上平台"Teachers Pay Teachers",为需要买卖教案和其他教育资源的老师提供了一个在线交易市场。[42]

虽然按需经济可能增加劳动力市场的参与人数,但是也会影响工作条件和劳工保护,人们对此也不无忧虑,例如:

- 由于按需经济属于一种新兴的经济模式,现有的关于劳工保护的法律法规,包括童工保护法和最低工资要求等,需要扩大适用范围。虽然部分按需数字平台提供劳工保护,但也有一些平台认为,劳动保护,即使是最基础的劳工保护措施,也不适用于按需劳动模式。

- 按需经济模式下工作场所无边界限制的特点,加剧了与工资和全球劳动力分配相关的问题。由于全球各地的生活成本不同,在用人单位能够在工资水平较低的地区聘请劳动者的情况下,未来许多职业都可能从工资水平较高的国家流向工资水平较低的国家。

- 部分研究表明,按需经济的受益人群主要是平台所有者和消费者,劳动者并未从中获益。[43]由于这些平台将工作降格为任务出售,因此可能低估劳动者对平台或数字经济作出的其他贡献。

- 劳动力商品化,也可能减少劳动者获得社会保险、职业发展和社会交流的机会,而这几个方面都能够加强创新、增加经济价值。此外，对于企业为创建工作文化作出的投资，按需经济模式下的劳动者也无法从中获益。

▸ 从长远来看，随着平台不断向劳动者"学习"，并且能够自动化处理的任务越来越多，平台经济的发展可能会使某些职业消失。无法获得新技能的劳动者可能会被边缘化，进一步使财富集中在平台所有者和高收入人群手中。

随着按需经济不断发展，企业有机会从公司内部出发，影响行业政策和公共政策。技术行业需要扭转其认知，既要从技术进步中获益，又要避免以员工失业、不受保护、无法获得福利或丧失长期职业规划为代价。

企业必须承认按需模式对劳动者产生的影响，而不应声称其"仅作为技术平台存在"。企业如果不承认劳工保护和员工福利的重要性，很可能破坏其品牌形象，并导致立法者和法院制定并实施监管措施，从而限制按需经济带来的商业机遇。微软认为，企业采取相关措施，为员工提供保护、福利和获得经济长期稳定的机会，并不妨碍其从按需经济中获益。

技术进步不仅支持了按需经济的发展，也在不断改变企业组织传统劳动力开展工作的方式。现在，许多企业都在打造全球分布式劳动力，其中的驱动因素有许多，包括需要在当地人才库之外，寻找具备企业所需技能的人员。另一方面,由于各国面临民族保护主义的压力,且企业需要遵守的移民法律越来越严格,因此,企业也需要考虑扩大国内的劳动力市场。

新技术和工具为企业提供了兼容分布式劳动力的能力。线上平台可以在不同的国家和地区收集有关劳动者和职位空缺的数据,轻松解决因地理位置造成的劳动者技能与空缺职位不匹配的问题。由于新的合作工具支持远程工作,因此,雇员不再受固定的工作地点约束。此外,劳动者也在寻找更加灵活的工作方式和工作地点。在最近的一次调查中,37%的技术类专业人士表示,如果能在家里办公,他们愿意少拿10%的工资。[44]

新技术在使企业有能力在全球范围内分配工作的同时,也要求企业转变培训员工、创建企业文化、建立机构知识和创造知识产权的方式。现在,许多企业发现,劳动力越分散,开展有效合作、保持企业灵活性就越困难。企业在建立新颖灵活的团队结构,将工作单元转变为任务制项目之后,需要重新思考聘用员工、建立团队、支持员工职业发展和培训的方式,将替代性雇佣安排和分布式劳动者相结合。例如,企业可能需要使用Microsoft Teams或Slack等合作工具,应对这些变化;使用LinkedIn Learning或Coursera等学习平台,满足员工对职业发展和指导的需求;此外,还需要探索新的方法和途径,在分散的劳动力中培育社群和敬业意识。

帮助每个人应对未来的工作挑战

在人工智能经济中,各种职业要求具备的技能变化速度极快,因此现有和未来劳动力的培养、教育、培训和再培训制度也务必随之演变。实际上,这不仅仅是新的人工智能经济的要求,人们早已逐渐意识到,在职业生涯中,大多数劳动者都需要不断学习新技能。[45]

皮尤研究中心最近的一份研究表明,在美国的劳动力中,87%的成年人表示,在职业生涯中不断接受培训并培养新技能,对于适应工作场所中发生的变化至关重要。为了在未来保持长期可雇佣性,劳动者所需具备的最重要的一项技能,或许就是学习新事物、合作、沟通和适应环境变化的能力。为使劳动者在变化迅速的劳动力市场保持竞争力,我们需要不断创新教育、培训和劳动力制度,并提出新的解决方案。

自动化和人工智能已经开始有能力执行需要思考和判断的任务,因此,为劳动者提供培训将越来越重要。我们需要重新关注人类的独有特质,培养劳动者的批判性思维、创造力、同理心和推理能力。

帮助教育和劳动力培训系统更好地理解、解释和预测用人单位需要的职业技能,是用人单位义不容辞的责任。虽然我们无法对未来的职业作出准确预测,但我们坚信,教育和培训会变得越来越重要。教育和劳动力培训系统可以更好地善用技术，帮助学生们和求职者发现有前途的职业生涯，评估其现有技能，培养新技能，找到对口职业，调整解决方案，满足不断增长的人口需求。

要成功适应自动化和人工智能时代，完善覆盖每一人的教育和培训制度是关键。大多数专家均同意，完善高等教育和培训是重中之重。图7和图8清楚地阐明了教育背景和就业水平的关系。图7表明，在经济合作与发展组织（OECD）成员国中，这两者存在正相关关系；图8表明，在美国，相较受教育程度较高的人群，失业率对受教育程度较低的人群的影响更加显著、强烈。教育背景不同，失业率增长比例也出现显著差异，且受教育程度较低的人群更易受到影响。这从另一个侧面证明了技术公司在制定教育和劳动政策方面能够发挥重要作用。

为了帮助人们获得培训，在现有的经济形势下立于不败之地，更好地迎接未来，微软正着手从以下三个方面努力：①帮助学生应对未来的职业挑战；②帮助劳动者应对不断变化的经济形势；③建立相关制度，更好地对劳动者和就业机会进行匹配。

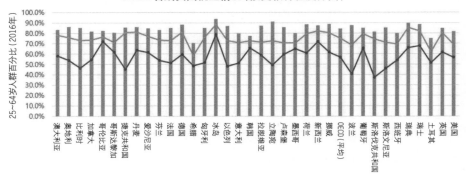

本图表明,在美国,由于劳动力市场对劳动者技能的要求更高,教育与可雇佣性之间的关系因此不断强化。

图7
来源:OECD,就业情况:按受教育程度分类,25~64岁人群百分比,2016年。

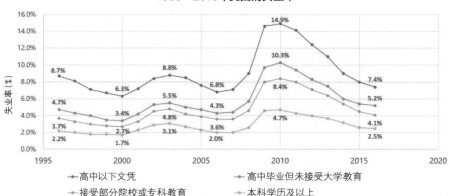

图8
来源:美国劳工统计局

帮助学生应对未来的职业挑战

要适应未来工作，人们需要掌握的最重要的一项技能，就是不断学习的能力。在未来，人类需要在工作中解决的问题将更加复杂，因此必须具备斯坦福大学教授卡罗尔·德韦克（Carol Dweck）所说的"成长型思维"。要取得成功，强大的沟通能力、团队协作能力和表达能力缺一不可。随着工作服务的对象不再限于一个社区，而将扩大为整个世界，因此，需要建立更加明确的全球意识。技术发展飞速，影响着各行各业，对数字技能——从基本的计算机素养，到高深的计算机专业知识——的要求将越来越高。新兴技术的发展，也会对人们的数字和计算机技能提出日益严格的要求。

鉴于这些需求在不断变化，年轻人在进入职场前需要学习的技能也发生了变化。除需要知道计算机如何工作、如何上网、如何使用生产力工具，以及如何确保计算机安全外，还应向他们提供学习计算机科学的机会。计算机科学培养的是一种计算思维，它实质上是解决问题的一种特殊方法，同时也是用人单位急需的一种技能。谁具备了所有这些技能，就能在高速发展领域获得高薪工作。因此，公平地提供严谨、有吸引力的计算机科学课程是首要任务。只有解决这一问题，才能为所有人群提供充分参与这一新职场的机会。我们最终的目标是打破种族、性别、生理缺陷、社会经济地位界限，打造多样化的计算机科学课堂。

部分国家,例如英国,已经在各年级都融入了计算思维教学,而其他许多国家仍在努力弥合数字技能和计算机科学教育上的差距。例如,虽然美国已经取得了重大进步,确保所有学生在高中毕业前至少可以选修一门计算机科学课,但仍有成千上万的学生没有机会修学这一课程。[46]根据美国大学理事会的数据,去年,美国37000所高中,只有4810所提供了大学预修计算机科学考试,其中女孩、少数族裔及经济困难者参与的机会最小。[47]

为了满足全球对发展数字技能的需求,微软公益事务参与了一系列项目和合作,旨在大规模地弥合这一技能差距。我们联手合作伙伴,努力帮助青年人为未来做准备,尤其是没有机会获得关键技能的青年人。例如,微软通过青年星火计划(YouthSpark),与来自60个国家的150个非营利组织合作,向300多万年轻人提供了校内外学习计算机科学知识的机会。

微软公益事务部与政府、教育机构、非营利组织及企业建立了合作关系,携手60个国家的150个非营利组织,为青年人提供校内外学习计算机知识的机会,迄今为止,我们已为300多万年轻人提供帮助,其中83%来自偏远贫困社区,且一半以上是女性。

要解决这一难题,增加计算机科学课的师资力量至关

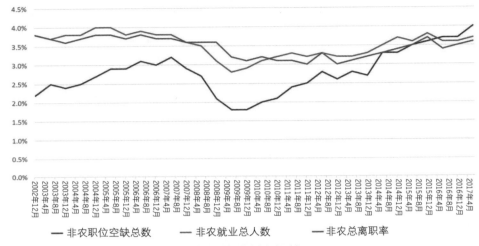

图9 来源：美国劳工统计局2017年10月发布的《职位空缺及劳动力流动调查》

重要。微软公益事务部支持的"学校技术教育和素养"项目（Technology Education and Literacy in Schools）已遍布美国29个州，入驻349所高中。来自500多家公司的1000名技术志愿者参与了这一项目，与数学或科学老师协作，为学生教授计算机科学知识。经过短短不到两年的时间，97%的任课老师便能够独立承担计算机科学课教学任务，为计算机科学项目的持续发展奠定了基础。

帮助劳动者应对不断变化的经济形势

技术变化的速度如此之快，仅注重对未来劳动者的教育是不够的，我们还必须帮助现有劳动者学习相关技能，应对不断变化的职场需求。具备高度技能的劳动者，是企业利用新一代数字技术创新的必要条件，也是推动经济增长的基础。为此，劳动者需要终生学习。如前所述，全球经济正在急剧变化，自动化和人工智能的发展，导致市场对具备高度技能的劳动者的需求越来越大。美国最新劳工统计报告——美国劳工统计局月度《职位空缺及劳动力流动调查》（JOLTS）显示，职位发布数首次超过了聘用人数。[48]

这只是一个例证，说明全球范围内用工需求与劳动者技能不匹配的矛盾已趋严峻。全球人力资源公司万宝盛华2017年的一项调查显示，在日本、印度、巴西、土耳其、墨西哥、希腊、澳大利亚和德国，技术人才缺口明显。[49]美国全国技工联盟报告称，虽然现今53%的工作是"中级技能"工作，或"新领阶层"工作，仅要求高中以上大学以下文凭，但符合这一要求的劳动力比例只有43%。同时，高中或以下学历的"低级技能"劳动者在整体劳动力中占比虽然高达20%，但只有15%的工作机会对其开放。[50]此外，Burning Glass技术公司发布的一项用工需求研究显示，每10份中级技能工作中，有8份要求具备基本的数字素养（digital literacy），但现在一半以上的劳动者缺乏该素养。除非我们改变教育和培训劳动者的方式，否则这一缺口将继续扩大。[51]全国技工联盟预测，到2024年创造的工作岗位中，有80%将需要中学以上文凭。[52]

随着对高学历、高技能劳动力需求的不断增长，我们必须找到新的途径来提高现有劳动者的技能。劳动力系统需要作出相应的发展，以跟上技术变化的步伐。善用新兴的远程教育和在线学习技术，加大对在职培训的投入，是使现有劳动者做好准备应对职场变化的关键手段。

要找到培训现有员工的正确路径，重点是明确企业所需的技能。微软及其子公司领英在这方面作出了一些新的尝试。[53]例如，领英联手美国国家网络安全中心（NCC）和位于科罗拉多斯普林斯的科罗拉多大学，确定了美国最热门的网络安全职位，并绘制出了这些岗位所需的技能。

领英还配合本土培训计划，更新了有关课程，并教授应届毕业生如何在求职过程中使用领英。微软也提供了一些课程和认证计划，帮助人们通过想象学院（Imagine Academy）、青年星火计划和领英学习平台（LinkedIn Learning）等渠道来培养数字技能。[54]事实上，国际数据公司（IDC）报告称，不论哪种岗位，掌握微软办公软件都是第三大热门技能要求。[55]

另外，寻找到新的劳动力市场切入点也很重要。由于人才短缺，企业必须探索出新的途径，从现有劳动力资源中挖掘人才。在这方面，微软和领英作出了多种尝试，微软软件和系统学院计划（MSSA）即是一例。MSSA是一个为期18周的培训项目，专门针对云开发、云管理、网络安全管理、数据库和商业情报管理类岗位，培训普通及资深员工。培训结束后，合格学员可以面试微软或合作用人单位的全职岗位。到目前为止，接收MSSA毕业学员的单位高达240家。此外，微软和华盛顿州学徒与培训委员会（Apprenticeship and Training Council）也有合作，后者推出了首个IT行业注册学徒项目。

领英也支持这种学徒模式，且正在设法打造一个学徒平台。领英推出了一个为期6个月的学徒项目，让学员加入领英的工程团队，了解软件工程师如何工作，累积相关经验，然后帮助他们走上软件开发岗位。此外，领英还联手CareerWise Colorado，创建了一个针对高中生的学徒招募平台。这一平台与科罗拉多州学徒办公室合作，增进了高中生对学徒价值的理解。

这些项目只是一个良好的开端，接下来更为艰巨的挑战在于如何通过公私合作推广这些项目，并对劳动者产生持续影响。这一方面要求培训机构全面转变培训思维，另一方面要求用人单位转变甄选和聘用人才的方式。

支持制定相关制度，构建以技能为主的市场

为促进全球经济繁荣，公共和私营部门还必须投资打造全新的教育模式。劳动者需要有机会习得市场所需技能，实现自我提升，同时也需要建立相应的制度，对相关资格认证

进行认可、汇总，并接受用人单位评价。职场的快速变化，要求用人单位和劳动者采用全新的合作方式。公共和私营部门应设法满足劳动力在各个发展阶段的需求，从初入职场的毕业生，到待业、半就业人员，再到需要习得新技能以确保长期就业能力的在职员工。

为了帮助企业找到合格员工和帮助劳动者找到工作，我们的制度需要从以传统学位为本转向以技能为本，说明各个岗位快速变化的技能需求，并指出劳动者个人应当拥有的技能，从而使劳动者更快地找到对口用人单位。

为此，第一步，要建立一个通用技能分类。新兴技术和职场变化要求教育机构提供用人单位需要的技能培训。在对劳动者开展相关技能培训，以及培训劳动者如何向潜在用人单位展示其技能之前，将市场最需要的技能加以整理形成条目，是非常重要的一项工作。用人单位和职介所应根据劳务市场实时信息来确定市场需要的技能。小至领英，大至IT行业，都可以很好地协助政府和职介所完成这一任务。政府可以根据这些信息制定和实施优质的劳动力培训计划，并对民营机构和非营利组织提供的培训给予奖励和经费支持。教育成果目标应涵盖就业、技能和发展方面的结果。

我们必须利用技术和数据建立一个动态的、以技能为本的劳动力市场，通过市场来引导教育和劳动力系统。要成功做到这点，我们需要建立一套以劳动者为核心的学习成果评估框架，不仅要统一各行业之间的数据，而且要方便个人查找。该框架应突出用人单位要求具备的知识，并涵盖劳动者在数字岗位上所需的专业技能和基础技能。基础技能包括解决问题的能力、职业道德、团队合作、好奇心和人际沟通。培训机构在帮助劳动者获得技能和证书的过程中，也应遵循这一框架的指导。

我们还需要确定现有的空缺职位和填补这些空缺所需的技能。领英、跑腿兔（TaskRabbit）、Upwork等数字平台，都根据职位或任务空缺发布了其对市场需求技能的看法。随着时间的推移，这类数据可用来构建领英经济图谱（LinkedIn Economic Graph）等分析资料，从而了解到特定技能的供求情况及其在特定地区随时间变化的情况——特别是，如果能够结合官方当地人口和企业数据情况，其作用将更加显著。

微软和领英还采取了一些其他措施，来了解哪些技能属于高需求技能，从而对技能培训进行投资，应对职场变化，帮助劳动者找到技能对口工作。为了实现这些目标，微软和领英与马克尔基金会（Markle Foundation）合作推出了Skillful项目。该项目旨在创建一个以技能为本的劳动力市场，面向所有人提供服务，但重点关注那些没有大学文凭的人。微软已投入大笔资金帮助马克尔基金会建立该市场。[56]

Skillful项目通过提供数据、工具和资源，简化了技能相关实践的采纳，为用人单位扩大了人才库。指导员和数字化服务能够为处于不同职业生涯阶段的求职者提供帮助，使之了解市场需要哪些技能，并获得相关专业培训。此外，项目还联手教育机构和用人单位，确保学生掌握制胜当代数字经济型社会所需的技能。该合作旨在建立一种可以在全美复制的模式，帮助数百万美国人找到满意的工作岗位。Skillful项目还联手领英，尝试通过各种策略，来提升高技能劳动者的生活水平；其中一个试点辅导项目是基于Skillful项目的指导服务和平台推出"导师通"（Mentor Connect）。

为了让公共部门和私营部门加强协作，更好地帮助求职者找到对口工作岗位，领英向美国各地政府免费提供了岗位展示服务。2017年，100多万个政府工作岗位出现在领英上。此外，全国州级职介所协会（National Association of State Workforce Agencies）下属的国家劳务交易所（National Labor Exchange）也于2017年1月开始向领英发送工作岗位，包括来自50个州岗位资料库的岗位。领英通过白宫技术人才招募项目（White House TechHire），与美国70多个城市共享对劳动力市场的洞见。领英还与纽约、洛杉矶、芝加哥、路易斯维尔、新奥尔良、西雅图、旧金山和克利夫兰的政府机构共享了有关数据，帮助他们更好地解决学生在校率、青年失业等问题，发现职业歧视及了解岗位技能的供求情况。

这些项目多数推出的时间并不长，但我们已经很明显地看到，我们需要利用数据建立一个更加动态的、以技能为本的劳动力市场，从而引导教育和劳动力系统，让劳动者做好求职准备。

改变规范，适应劳动者不断变化的需求

为了应对经济发展的挑战，我们还必须了解按需经济、兼职工作、独立承揽和临时性工作对个人和社会的影响。[57]这些变化引发的一些问题，可能无法通过现有法律和政策框架得到彻底解决。

为了实现创新和保护劳动者，公共和私营部门必须解决大量关键政策问题。必须制定明确的法律规定，使劳动者和企业清晰了解自身的权利与义务。行业内部也须制定自己的劳动者保护标准，避免社会贫富分化进一步加剧。为了促进技能有效流动，鼓励创业，让劳动者充分发挥其市场力量，行业和政府必须通力合作，设法使劳动者在换工作时，福利能够随迁，不致中断。社会保障体系必须完成现代化，为劳动者及其家属提供支持，并在经济动荡和劳动力市场波动期间，保持整体经济稳定。

向用人单位和劳动者提供法律确定性和法律架构

鉴于现代劳动力的变化速度如此之快,现有法律和政策框架无法充分应对今天的工作安排并不奇怪。劳动者分类的方式和不确定性成为问题已有一段时间,对企业、劳动者和政府都造成了影响。现在,劳动力市场的变化和按需工作平台的兴起,进一步加剧了寻找这些问题答案的紧迫性。[58]

广义而言,现行法律一般将劳动者分为两类:①受用人单位正式聘用提供长期服务的劳动者,被称为雇员;②按具体合约提供货物或服务的劳动者,被称为独立承揽人。

传统上,雇员在工作时间和工作环境上比较缺乏自由,随意性较低,但工作一般更加稳定,所受法律保护也更多;而独立承揽人在工作时间和工作方式上相对自由,但受到的法律保护较少。是否构成雇员,决定了他们是否受到传统劳动法、工资工时法及平等机会法保护,以及是否能获得用人单位提供的福利,如私人养老金、培训机会、退休金、多数国家都提供的医疗保健等;还决定了用人单位是否会承担一部分社会保障福利费用,以及雇员能否享受到社会保障福利,如失业保险、美国的社会保险和国家带薪休假津贴等。

如今,数字平台和用人单位将大多数按需劳动者视为独立承揽人。这种做法使得按需劳动者不受最低工资和加班费规定、童工法规及反歧视反骚扰法律的保护。此外,通过中介撮合的劳动者依法能享受到的权利和保护也不明确。由于对缺乏保护日益不满,按需劳动者越来越频繁地通过诉讼或政府干预来反抗这种做法。[59]

诉讼和政府干预的结果飘忽不定。对于按需劳动者来说,这导致他们无法确定地预期自己享有的权利和福利。对于平台公司和用人企业来说,这带来的问题是按需劳动者是否会被视为雇员,从而产生相关成本,并需要为其提供相关保护。

在劳动就业法和福利制度适应当前劳动力趋势,完成现代化之前,产能的提高和机遇的增加有受到遏制的危险。目前存在一种风险,即,如果我们未能强制实施基准保护(包括工资保护),那么工作岗位的两极分化将越来越严重,一极是高薪、稳定的岗位,另一极则是低价值、低薪、基于任务的临时岗位。这可能会损害按需经济的潜力。不幸的是,目前关于劳动者分类的观点往往是各执一词——企业强烈要求按狭义划分,而劳动者则强烈主张按广义划分。我们需要针对企业需求和劳动者需求开展更加广泛的对话,从而确定需要作出哪些改变,以一种富有成效和公平的方式满足双方的利益。

到目前为止,政策建议不外两种倾向:一是重新定义雇员和独立承揽人这两种劳动者类别,二是设法减轻这种差异造成的后果——主要是向临时工提供保护、福利和社会保障体系参与权。两种方案都是想通过弱化二者差异来解决问题，为目前未被覆盖的劳动者提供基本保护。当前的政策建议包括在雇员与独立承包人之间，确立"独立劳动者"这一全新分类；为某些劳动者建立一个所得税和就业税的纳税避风港；向某些类别的按需劳动者提供劳资谈判等保护；通过关于劳动者保护的自主最低行业标准。鉴于越来越多的人通过按需平台找工作，所有这些建议都有探讨的必要。

制定行业标准，保护所有劳动者

在今天，商界领袖可以制定自己的按需雇佣标准，从而有机会发挥重大作用，重塑新兴经济下的雇佣政策。微软认为，我们可以（也应该）通过内部政策积极影响按需劳动者的待遇。微软制定了适用于所有按需工作的最低工资规定。根据微软的政策，按需劳动者应在完成工作后一周内获付报酬，所有劳动者都应保有尊严、获得尊重。微软同时禁止使用童工，并要求其使用的按需平台具备可及性。微软正在与其使用的按需平台落实体现上述政策的合同条款。

虽然企业政策可以向按需劳动者提供一定程度的保护，但其影响是有限的。使用按需劳动者的企业也可以参与、促成更广泛的解决方案，来解决这类问题。例如，自由职业者联盟和护工联盟等团体完善了任务劳动者标准——部分标准甚至是通过立法来完善的。全国家政劳动者联盟（National Domestic Workers Alliance）针对美国家政劳动者发布了一份《好工作准则》（Good Work Code），提供了一个雇佣劳动者的框架，涉及安全、共享美好生活、生活工资、包容和投入等方面。[60]行业领袖应该鼓励企业和劳动者之间开展讨论，制定类似的任务型工作标准，内容可以涵盖工资、福利和公平待遇承诺等。这可能会导致一系列标准的出台，不仅可能得到企业认可，还可能作为非官方政策框架。这类标准可以是适用于特定行业的标准，也可以是针对更广泛平台的普适性标准，同时还可以在立法确立最低保护要求时，被用作基础框架。

确保福利随迁

上述劳动力市场趋势对劳动者保护和用人单位提供的福利均有巨大影响。以用人单位为核心的福利制度在20世纪中叶出现在欧洲和北美大部分地区，其基础原理有二：第一，健康稳定的劳动力给企业创造了效益；第二，为维持劳动力稳定，某些福利最好由用人单位（而非政府）提供。

这种做法形成了我们对劳资双方社会契约关系的刻板印象。虽然工作性质随技术创新发生了变化，但企业福利制度和社会保障体系却停滞不前。我们目前面临的挑战是，如何改革福利和社会保险制度，为劳动者提供充分保障，同时为企业建立一种长期可行的出资机制。

在当代数字经济社会里，劳动力的机动性，以及迅速将人才整合到新增长领域的能力，对企业成功至关重要。许多企业可能会发现，维持福利往往负担较重，代价过高。个人劳动者也希望福利能够随迁且具备弹性。福利的随迁性，对有效解决问题至关重要。目前存在三种可能的解决办法。

▶ **由用人单位提供福利**。在以短期项目为主的行业，向劳动者提供福利并不鲜见。建筑业、娱乐业等行业通过劳资合作关系（labor-management partnerships）解决了这一问题。通过这一策略，劳动者在多个用人单位之间流动时，仍能保留医疗和养老保险，从事短期工作也无妨碍。劳资谈判机制为用人单位提供了一个途径，使其可以无需承担行政责任，直接对福利池进行出资；劳动者不必负责迁移福利和寻找新的福利提供者。新模式可以使用本办法，这将减少低效和混乱问题，确保劳动者获得基本的保护和充分的福利。这种做法会导致劳动力流动更频繁，因为劳动者仅仅为了保住福利而留在原工作岗位的可能性将进一步降低。

▶ **使用新平台提供福利**。按需劳动力平台的兴起，可能会创造机会，为劳动者提供获得福利的新途径。例如，护工平台Care.com采用类似于传统企业为员工福利出资的方式，让家庭为护工福利提供出资。[61]当家庭通过Care.

com支付护工费用时，资金会按一定比例形成福利，即使护工改为服务Care.com上的其他家庭，福利也会延续。但本办法仍然面临一些挑战，比如劳动者为不同平台的客户服务时的福利处理问题。

- **政府指令和资金**。在部分国家，国家政府组织甚至是跨国政府组织可能会设法填补这一缺口。在无法在全国范围内推行新制度的国家，或许可以由较小的政府部门建立负担福利所需的基础设施和风险池。有些国家需要建立基本的福利制度，和发放这些福利的配套机制。以美国为例，若干新的联邦宏观项目没有得到政府支持，但一些州设法建立了自己的医疗或退休计划。从短期来看，政策制定者应该考虑建立试点项目，打造可迁移福利，比如美国各州推出的法律法规等。[62]

建立现代化的社会保障体系

高流动、高活力的劳动力会加大社会保障体系的压力。随着人们开始通过更多样化的非排他性安排获得工作，而这类安排可能并不像用人单位那样提供福利，或者无法让劳动者挣到足够的金钱作为积蓄，他们将比以往任何时候都更加依赖失业保险、工伤补偿、社会保险等社会保障体系。

在劳动者工龄期启动的保障体系对劳动者的经济稳定尤

为重要，经济稳定反过来又有助于维持多样化的高技能劳动力队伍。在无活可干的期间，劳动者会产生收入波动，从而对劳动者及其家庭成员产生严重的长期影响，同时也会减少企业可用的高技能劳动力储备。即使经济发展强劲时，失业或半失业水平也很高。2017年8月，美国劳工统计局估计，美国有710万劳动者失业，另有530万人因经济原因兼职或被迫兼职。在劳动者的一生中，这种期间可能多次反复出现。[63]

许多现有社会保障计划已经出现资金投入不足问题，而且，随着劳动力逐渐老龄化，其面临的财政压力将越来越大。这意味着，在经济衰退期等对社会保障需求增加的时期，现有保障计划可能不具备相应的应对能力。雪上加霜的是，许多综合性保障计划严重依赖传统雇佣关系。如果传统就业格局已发生重大转变，而政策没有相应变化，可能会进一步侵蚀以传统工作岗位为中心的社会保障计划。最后，这类计划没有考虑到新的工作模式，也没有预料到个人可能会因各种原因而提高换工作的频率。要鼓励劳动力流动，使劳动者能够习得新技能、获取新机会，建立现代化的社会保障体系是重中之重。

公司可以探索通过公私合作关系来满足劳动者的需求。例如，微软正在通过领英探索加速美国劳动者再就业的新途径。领英联手犹他州试验开展了一个网络求职试点项目，试图加速失业者的再就业过程，节省纳税人的资金，这个项目得到了特朗普政府的特别重视。此外，微软和领英还为就业顾问和求职者打造了系列工具，以期完善失业保险、各州劳动力保障计划等劳动力相关计划。领英还联手全国州级职介所协会，为其在美国的2500家公共托管就业中心制作了求职课程。

企业应继续利用数据和技术工具，协助政府发现重新安置劳动者的机会，并在试点成功后推广这些解决方案。但是，建设现代化的社会保障体系需要采取多元化的措施，如：

▶ **重新审视失业保险和再就业计划，包括职业培训和转行补助项目。** 已有相关提议建议采取相关措施，建立现代化的失业保险保障计划，并增强这一计划的偿付能力。企业应考虑新的工作模式，参与讨论建立全新失业保险和就业服务制度的重要性，应预料到劳动者更频繁换工作的可能性，还应促进劳动力流动，帮助劳动者习得新技能、获取新机会。

▶ **改革税收政策和社会保障体系。** 政策制定者必须探索如何调整政策，为社会保障体系计划提供充足资金，包括在现有税基以外，考虑其他提供资金的方法。例如，有人质疑，工资是不是衡量应税收入的正确基准。既然企业生产力通过产量而非工资来衡量更好，有人建议通过其他标准来征税，以支撑社会保障体系和政府收入。

同样要探索的是，社保计划如何才能扩大劳动力规模，如何架构才能帮助雇员更轻松、灵活地换工作，以及减轻用人单位的负担。社会保障体系在没有进行充分现代化的情况下，将无法为新兴工作模式提供充分支持。私营和公共部门必须联合起来，探索如何在新经济中为劳动者提供最佳的支持。

通力合作

在前进的过程中，政府、私营部门、学术界和社会部门应联手探索如何在新经济中为劳动者提供最好的支持。这一目标可以通过以下手段实现：开发新的培训和教育方法，以便在技术进步时，劳动者能习得用人单位所需的技能；以创新方式为劳动者提供工作机会；建立现代化的劳动者保护机制，促进劳动力流动，并降低全球经济快速发展给劳动者及其家庭带来的不确定性。

结论

人工智能
增强人类创造力

> 结论
> 人工智能
> 增强人类创造力

一旦计算机的计算智力被用于提高人类的智慧和创造力，会发生什么奇妙的化学反应？何谓以人为本的人工智能？

我们或许可以从Melisha Ghimere身上，窥见这一问题的答案。Melisha Ghimere今年20岁，是尼泊尔加德满都坎提普尔工程学院计算机科学专业的学生。在2016年举办的微软创新杯大赛中，Melisha Ghimere的团队打入了区域决赛。

与大多数尼泊尔人一样，Melisha Ghimere来自一个自给自足的农牧家庭，他们养殖奶牛、山羊和水牛。Melisha Ghimere的叔叔和婶婶Rajesh和Sharadha经过经年累月的辛勤劳作，成功使牧群巅峰时期的规模超过了40头牲畜，不仅足以养活两个孩子，还可以给其他四个亲戚提供帮助，甚至雇用了几名帮工。但是，7年前，炭疽病疫情爆发，Melisha Ghimere的叔叔和婶婶丧失了大部分牲畜，至今，他们仍未走出这场灾难的阴影。

就读大学期间，Melisha从未忘记她的家人。为了帮助和她叔叔一样的牧民，她着手开发一项技术解决方案。Melisha与其他三名同学共同研究畜牧和兽医实践，深入采访牧民，最后成功创建了一个监控装置原型，用于跟踪家畜体温、睡眠模式、焦虑水平、身体移动和活动。这一人工智能系统可观察每头牲畜的微妙变化，并基于观测数据，预测牲畜的健康状况。牧民可通过移动电话跟踪牲畜的健康状况，获取有关牲畜健康的建议，并在牲畜出现疾病、焦虑症状或可能怀孕时收到相关提示。

Melisha的项目还刚刚起步，但初步成果已颇令人鼓舞。在首批实地测试中，牲畜健康状况预测的准确率高达95%。已有一户牧民借助这一解决方案，在还没有意识到牲畜疫症的情况下，检测出一头处于炭疽病感染早期阶段的奶牛，从而避免了疫情爆发可能导致的大规模牲畜死亡。

与Melisha的项目一样，人工智能本身也还处于初始阶段。经过过去几年的努力，我们开始建造具备感知、学习和思考能力，并可以在此基础上提供预测或建议的系统。几乎每个人类活动领域都可获益于为配合人类智慧而设计的人工智能系统。从防范致命疾病，到帮助有残疾的人更全面地参与社会生活，到创造更加可持续的方式来利用地球的有限资源等，人工智能为我们描绘了一个更加美好的未来。

结论
人工智能增强人类创造力

如此规模的巨变，不可避免会引发一些社会问题。在计算机时代，我们也曾被迫应对许多重要的问题，例如隐私保护、安全、可靠、公正、包容，以及人类劳动的重要性和价值。随着人工智能系统的用途和应用越来越广泛，所有这些问题的重要性都将进一步凸显。为了确保人工智能能够用于我们所期望的目的，我们必须找到正确的答案，且这一答案必须全面理解和满足人类的愿望、需求、期待和需要。

这就需要对人工智能采取以人为本的态度，体现人类永恒的价值观，还需要坚定不移地秉承驾驭计算智能造福人类的主旨。我们的目的不是用机器替代人，而是利用人工智能通过分析大规模数据最终找到那些难以发掘的规律那样的无与伦比的能力，来增强人类的能力。

我们无法预测人工智能将对我们以及我们后代的生活造成什么改变。但是我们已经看到，Melisha制作的这个检测装置，可以帮助偏远地区数以百万计的小牧民过上更好的生活，从中，我们也看到了一个利用人工智能提高人类的智慧和想象力，改变人类生活的例证。

我们相信，世界上有着数以百万计的像Melisha一样的人，他们年龄或大或小，但都对如何利用人工智能应对社会挑战充满想象力。试想一下，如果这些人能够获得人工智能所提供的各种工具和能力，他们将产生多少洞见，解决多少问题，做出多少创新。

当然，这一切不会从天而降。只有研究人员、政策制定者、政府领导人、企业和公民社会同心协力，携手为人工智能制订一个可共享的道德标准框架，才能真正实现以人为本的人工智能。同时，可共享的道德标准框架也有助于以负责任的态度开发人工智能系统，并增强对它的信任。在进一步向前发展的过程中，我们期待与各行各业、各个领域的从业者共同合作，开发和分享最佳实践，为人人信任、以人为本的人工智能奠定坚实的基础。

补记

人工智能的
中国使命

补记
人工智能的中国使命

今天的中国正在全球的人工智能舞台上迅速崛起，中国在全球人工智能产业的影响力与日俱增。我们看到越来越多的全球人工智能专家正在向中国集聚，不论是来自微软的人工智能专家们，还是来自其他公司或是高校和科研机构的专家与学者们，都在赶赴中国的人工智能"盛宴"。

从全球角度来看，中国近年来在人工智能的技术和产业推进方面，具有独特的优势。中国有着全球最大规模的互联网、移动互联网和智能手机用户数量。根据中国互联网络信息中心（CNNIC）的统计数据，截止到2017年12月，中国网民规模达到7.72亿，中国网民普及率达到55.8%，超过全球平均水平（51.7%）4.1个百分点、超过亚洲平均水平（46.7%）9.1个百分点，手机网民规模达7.53亿，网民中使用手机上网人群的占比由2016年的95.1%提升至97.5%。

计算未来

我们知道，人工智能的普及以及技术进步有赖于两大因素，一是可用数据的规模，二是技术用户的数量。而这两点在今天的中国，都有着全球其他国家和地区所无法比拟的优势。以中国网民规模而言，根据CNNIC，到2016年底就已经达7.31亿，这相当于欧洲的人口总量。而就数据总量而言，中国由互联网和移动互联网所产生的数据量，已经形成规模化优势。在中国的"双十一"购物节中，仅11月11日当天就能够产生超过十亿笔的在线支付、超过20万笔/秒的支持峰值以及近十亿的物流单（2017年）。

在中国政府出台的《新一代人工智能发展规划》中，明确指出，人工智能将催生新技术、新产品、新产业、新业态、新模式，引发经济结构重大变革，深刻改变人类生产生活方式和思维模式，实现社会生产力的整体跃升。今天，这样的改变正在中国社会开始发生，例如今日头条、京东、天猫这样的移动应用正在主导社会的注意力，从而大规模改变或影响信息流、物流、资金流等社会基础资源的流向，新零售、新制造、新金融等新业态涌现。

正是因为中国政府对人工智能的高度重视，才有了今天人工智能的中国机遇，也同时带来了人工智能的中国责任。一方面，中国要为全球人工智能发展贡献中国智慧——根据CNNIC统计，截至2017年6月，中国拥有人工智能企业592家，占全球总数的23.3%，2016年中国人工智能相关专利申请数达30115项。另一方面，中国也要考虑人工智能算法的社会责任——如何避免人工智能对于社会基础资源再分配与导向的不公平、不公正、缺乏包容等影响，如何通过人工智能让更多的人参与到社会的发展机会中，如何让每个个体与组织在人工智能的平台上成就更多。

中国有一句话叫做：天时、地利、人和，现在中国从中央政府到地方政府再到普通老百姓，都已经开始全面拥抱人工智能。如今，正是需要中国探讨如何加深对人工智能的信任，从而让人工智能在中国的社会经济中发挥更大作用的时机，在抓住机遇的同时，承担应有的责任。

中国政府也有意愿和态度加强在人工智能政策法规、伦理道德等方面的研究。《新一代人工智能发展规划》提出在人工智能的发展环境方面：到2020年要初步建立部分领域的人工智能伦理规范和政策法规；到2025年要初步建立人工智能法律法规、伦理规范和政策体系，形成人工智能安全评估和管控能力；到2030年要建成更加完善的人工智能法律法规、伦理规范和政策体系。

微软作为一家全球性的公司，鼓励各方广泛对话和持续合作，积极发展以人为本的人工智能，塑造人工智能的未来，最大限度地发挥其潜力，减轻其风险和影响。在建立对于人工智能的信任方面，中国与全球其他地区都处于同一个起跑线上。无论是中国大力进行的数字化转型,或是中国经济在全球的快速发展,中国都有意愿、有能力参与全球化数字经济发展、贡献中国智慧、承担中国责任,这就是人工智能的中国使命,也是微软的使命。

数字经济——全球化普惠机遇

云计算、大数据、人工智能等新技术正在激活一个全新的经济形态:数字经济。中国信息通信研究院认为,数字经济是以数字化的知识和信息为关键生产要素,以数字技术创新为核心驱动力,以现代信息网络为重要载体,通过数字技术与实体经济深度融合,不断提高传统产业数字化、智能化水平,加速重构经济发展与政府治理模式的一系列经济活动。

今天,数字经济已经成为中国转换经济增长动能的新方向,中国正进入全面实现数字化转型、大力发展数字经济的新时期。2017年,"数字经济"一词首次进入了中国政府工作报告,提出"将促进数字经济加快成长,让企业广泛受益、群众普遍受惠"。作为全球第二大经济体,中国经济的数字化转型以及发展数字经济,也将对全球数字经济的进程产生深远的影响。特别是中国承载了全球近五分之一的人口,人民生活即将全面实现小康。因此,数字经济的发展有助于提高这全球五分之一人口的普惠发展机会。

微软认为，中国的数字化转型带来的不仅是更高效、更智能的生产力工具，更重要的是通过数字化转型为每一个组织和每一个个人带来普惠发展，能在这场变革中，成就不凡的未来。微软与市场调研公司IDC联合发布的一份报告显示：到2021年时，数字化转型将为中国GDP的年度增长贡献1%的增速，价值7160亿美元。在这样一场7160亿美元的商业"盛宴"中，每一个人和组织都有机会通过数字技术平台和工具，参与其中。

而人工智能可以说是数字化转型和数字经济的发动机和动力引擎，特别是随着承载人工智能的公有云服务越来越廉价、覆盖面和地区越来越广泛、可靠性和安全性越来越高，只需要一部手机就能接入人工智能服务。而人工智能能够与边缘计算和云计算很好地结合，用于生产数字化产品与服务：训练好的人工智能模型可以自动化地在物联网、智能手机、智能工业设备等数据源附近就近实时计算，也可以在云端通过云计算数据中心完成大规模的大数据批量计算，在网络通信带宽良好的前提下也可以在云端完成大规模的实时数据计算，这些自动化的人工智能计算能够源源不断地把数据转换成为洞察，也可用于提供智能化服务。

正是这种自动化、低成本、无处不在的人工智能计算，为中国的每一个组织和每一个人，带来了广泛受益、普遍受惠的发展机遇。

智在中国，惠及全球

计算未来

中国在积极参与由人工智能驱动的数字经济大潮时，也为全球人工智能的发展提供了独特的场景与实验田。来自中国的人工智能研究成果，也借由全球华人AI社区向全球扩散。来自领英的AI人才报告显示，截止到2017年第一季度，全球拥有人工智能领域专业技术人才超过190万人，其中美国人工智能领域专业技术人才总数超过85万人，中国人工智能领域专业技术人才总数超过5万人，全球共有近14万华人AI技术人才。

中国是微软在全球范围内最重要的创新中心之一，在积极与中国同步共享全球领先技术的同时，微软也在汇聚中国的本土智慧，造福整个世界。微软亚太研发集团是微软在美国之外的最大研发组织，拥有众多华人AI人才，也创造了举世瞩目的研究成果。今天，微软亚太研发集团的研究成果，正通过微软的平台向全球扩散。

微软"小冰"是在中国土生土长起来的人工智能技术。2014年，位于北京的微软（亚洲）互联网工程院研发出了初代"小冰"。微软"小冰"最开始被塑造成一个"16岁"可爱萌妹子形象，在多个社交平台上与人类对话。在情感计算框架的基础上，通过人工智能算法、云计算和大数据的综合运用，采用代际升级的方式，"小冰"逐步形成向EQ情感智能方向发展的完整人工智能体系。

我们首倡"人工智能创造三原则"，而微软"小冰"正是这三原则的最好体现：

- 人工智能创造的主体，须是兼具IQ与EQ的综合体，而不仅仅是具有IQ；
- 人工智能创造的产物，须能成为具有独立知识产权的作品，而不仅仅是某种技术中间状态的成果；
- 人工智能创造的过程，须对应人类某种富有创造力的行为，而不是对人类劳动的简单替代（如工业机械臂那样的"人工智能制造"）。

微软"小冰"通过理解对话的语境与语义，实现了超越简单人机问答的具有EQ的人工智能体验，特别是包容了中国当代文化，尤其是具有当代中文互联网文化的人工智能情感体验。2017年5月，微软"小冰"甚至推出了原创诗集《阳光失了玻璃窗》，这也是人类历史上第一部100%由人工智能创作的诗集。微软"小冰"从1920年以来的519位中国现代诗人的作品中学习，经过超过1万次的迭代学习，逐渐形成了自己风格的诗作。

> 树影压在秋天的报纸上
> 中间隔着一片梦幻的海洋
> 我凝视着一池湖水的天空
> 我们来到这个世界
> ……
> ——摘自《阳光失了玻璃窗》

此外，微软"小冰"在有声少儿读物方面，其质量超越98%的人类创造者。而经过测试，将少儿读物转变成有声读物的用时，"小冰"仅为同水平人类的1/500，成本仅为同水平人类的1/80000。目前，通过微软"小冰"大规模生产的有声读物已经投入市场。2017年12月3日，微软"小冰"获得2017年世界互联网领先科技成果奖。在2018年4月的博鳌亚洲论坛上微软"小冰"与来自全球的嘉宾互动，让他们全方位体验源自中国的人工智能产品，受到好评。

如今，微软"小冰"已经走向全球。微软"小冰"的全球化策略是优先选择人口数量超过1亿的国家，在当地建立完全本地化的团队，从而确保"小冰"根植于该国本土文化。2015年及2016年，微软分别推出日本"小冰"和美国"小冰"，2017年2月和8月又推出了印度"小冰"和印度尼西亚"小冰"。截至2017年8月，微软"小冰"拥有了超过1亿人类用户，对话量超过300亿轮。在全球可以体验微软"小冰"的移动互联网平台，有Facebook Messenger、LINE、微信、QQ、Windows 10、美拍、京东、米聊、米家、优酷等。

能力有多大，责任就有多大

得益于人工智能，今天中国几乎在所有领域都在以更快的速度、更大的步幅前进，我们应该同时注意到：

人工智能如何助力消除贫富差距？掌握技术的人群通常会更多地享受到技术带来的益处，而大部分人尤其是低教育人群将会在贫困中而无力改变现状。通过人工智能技术研发，推出的更多易用、易得、普适的人工智能产品将助力改变这一状况。特别在中国，如何通过人工智能提升农村地区、老少边穷地区等低教育人群参与新经济的机会。此外，人工智能将取代一部分现有的工作，同时也会产生新的技术职位。对于由于技术变革而造成失业的人群，我们该如何妥善安置。这需要政府利用制度予以保障，比如通过税收、福利、全民普遍基本收入（UBI）等政策。

人工智能技术如何确保人身和财产安全，对人类的隐私和其他基本权利不造成威胁？产品设计者和技术的拥有者应该负有什么样的责任。比如，如何确保无人驾驶汽车不会失控，如果失控带来的责任应该由谁来承担。

人工智能造成的偏见和垄断如何解决？比如根据用户的偏好，进而选择性地推荐、筛选内容，造成用户认知上的偏见。技术上的垄断，造成了一些科技产品成为了类似水、电这样人们生活所必需的基本资源；而一些社交应用已经成为人们日常交流的必需品，垄断了人与人之间的关系。中国是否应该对类似的技术垄断加以管控，对有可能引起或者加剧民众偏见形成的技术给予重视？

对于上述人工智能的社会责任，微软认为应该通过对话的方式加以应对，但最终也必须要落实到一个可信、可靠、安全的人工智能技术平台上，才是最终的解决之道。由世纪互联蓝云运营的微软Azure智能云平台本身在进入中国市场的时候，就针对中国市场的法律法规与合规环境进行了调整和优化，我们还提供灵活的接口以及接入方式，以持续兼容新的监管规则。

有中国特色的 AI 之路

鉴于人工智能对于中国数字化转型、发展数字经济的重要性，在遵循微软六大人工智能价值观的前提之下，微软与合作伙伴也探索了有中国特色的AI之路。今天，基于微软六大人工智能价值观的微软智能云平台正在全面驱动中国的数字化转型。

数字码头打造未来海运的"中枢神经"

作为全球最大的港口起重机制造商，上海振华重工集团在全球航运业的地位举足轻重。目前，振华重工占有全球港口设备约80%的市场份额，有97个国家、超过250个港口正在使用振华生产的起重机和重型机械。振华重工正在从传统的制造业，转型为新一代数字化的智能港口服务提供商。振华重工利用微软云打造物联网平台，连接设备、分析实时数据，并汇集到全球的监控中心。振华重工集团董事长宋海良表示："我们原来卖硬件，现在我们卖软件、卖服务。在这个变革的路径中，振华作为全球港口制造企业的领头羊，一个国际化的公司，与微软公司合作，创造新的产品、新的服务和新的模式。"

深度学习结合时空数据，让城市管理者对人流量"未卜先知"

近些年，几起由于人流量激增引起的公共踩踏事件引发了政府和社会对城市交通突发事件的关注。踩踏事故的频繁发生，让研究员们下定决心要解决城市人流预测问题。过去的人流预测往往是从预测个人行为的方式着手，但对于大规模人流预测，始终难以应对。这是由于人流数据具有时空属性，影响因素非常多。微软亚洲研究院从时空数据出发，设计出一个特殊的网络模型：将城市分割成均匀的网格，输入人流数据（包括手机、出租车轨迹等），来计算出每个网格的人流进入和流出量，之后再对时间特性以及空间属性进行模拟，并通过引入深度残差网络结构提高训练精度，得出更

精准的数据结果。2016年,微软亚洲研究院的人流量预测系统就已在贵阳市进行了落地实验。

"云"中漫步,让出行更便捷

随着共享单车行业的爆发,各大共享单车企业的发展步伐再次提速。作为行业中的明星企业,摩拜不仅进驻国内各大城市,同时作为全球最大的智能共享单车平台,也走进了全球130多个城市。随着业务的扩大,背后的单车数量、用户数量、用户使用数据、管理数据等数据处理压力,也以指数级别攀升。为应对这一挑战,摩拜将其整个数据平台都迁移到了微软智能云Azure上。借助Azure智能云平台可灵活扩展、安全可靠的特性,摩拜实现了对全球几百万辆单车和海量用户数据的高效管理。 基于Azure IoT物联网服务、Dynamics 和CRM运营,以及机器学习的预测分析和智能服务的帮助,摩拜可以提前预测什么时候挪车、挪车数量、路径优化,并运用大数据分析,提升每辆单车的日使用量,降低运营成本。

智能医疗升级,Airdoc 让更多人拥抱光明

中国每年都有数以万计的糖尿病患者因为其并发症——糖网病(糖尿病性视网膜眼底病变)的治疗不及时而失明。

但也有众多糖尿病患者及时采取有效措施,挽救了双眼视力。作为一家专注于以深度学习提升医学诊疗效率的成长型企业,Airdoc帮助医生提高诊断效率和准确率,尤其是对于糖网病来说,可以协助医生快速完成对视网膜病变的筛查及分析,让医生拥有对糖网的识别和诊断能力,提高糖网病管理能力。而Airdoc就是借助微软认知工具包,解决了自身所面临的大规模医疗数据处理和实时并发需求。

智能云助力全国农村学生营养改善计划实施

2011年底,中国国务院启动"农村义务教育学生营养改善计划",由教育部牵头实施,每年投入180～200亿元,为29个省、699个贫困县的超过13.4万所农村义务教育学校、3200多万名中小学生提供营养膳食补助,帮助改善农村学生营养状况,促进教育公平。中国发展研究基金会与微软达成合作,搭建了基于微软智能云Azure的"阳光校餐"数据平台,结合互联网、大数据等技术,实时监测"农村义务教育学生营养改善计划"政策落实的情况,并提供数据可视化分析,客观、科学地反映执行效果,为政策建议提供可靠的数据依据。目前"阳光校餐"数据平台已覆盖超过1万所学校。"阳光校餐"实施以来,最大的受益者是因为补充了营养而健康成长的贫困地区儿童。随着营养改善计划深入细致地执行,更多的孩子们可以吃得更营养,长得更健康。

助力STEM教育，让想象力生长

创建于2013年的深圳市创客工场科技有限公司（Makeblock），希望能降低创造的门槛，让更多人能享受创造的乐趣。随着海内外STEM（科学、技术、工程及数学四门学科的英文首字母缩写）教育市场的逐步发展，Makeblock团队开始聚焦将技术和教育有机结合，研发适合STEM教育市场的产品，打造软件、硬件、课程与机器人赛事相结合的健全产品线，希望增加孩子们学习机器人编程等知识的兴趣。如今，Makeblock的用户已经遍布全球140多个国家，也分别在美国、荷兰及日本设立子公司，用户数量超过200万。在Makeblock机器人和图形化编程产品中，微软认知服务为其提供人脸、情绪识别等人工智能类功能。微软认知服务帮助Makeblock在行业中增加了认知度和接受度，不仅让AI走进更多的学校和家庭，也让更多的孩子和创客受益于微软认知服务的计算能力，让想象力得到最大的释放。

人脸识别技术助力"宝贝回家"

微软人脸识别API是一项基于微软智能云的服务，它可以对人脸图像进行扫描，利用先进算法比对两张照片的相似度。在寻找走失儿童的过程中，基于微软人脸识别API的应用程序"回家"能够分析27个人脸面部特征，因此即使拍摄角度不同、面部表情各异、或者跨年龄比对，该应用也能从诸多张照片中准确识别面部相似的图片。通过和公益机构"宝贝回家"的合作，微软为中国最大的公益寻亲网站宝贝回家寻子网提供人脸识别的技术支持，为数以万计的走失儿童照亮回家之路。

智能物联网让盲人听见巴士

"听听巴士"是一款帮助视障人士便捷出行的APP。2016年9月，作为广东十大民生工程之一的公交导盲项目，广州3000辆公交车上装上了导盲终端。视障人士要乘车时，打开APP后摇一摇手机，APP便会自动告知身边的站台以及该站台可以乘坐的公交线路，在站台候车时还能听到"公交车还有几站到站"等信息。这一项目的实施落地，给盲人出行带来了极大的便利。"听听巴士"的成功，离不开物联网标签的大规模应用，随着平台所支持的运营线路和用户的增长，"听听巴士"车载终端和物联网标签的数量也呈现指数级别增长。Azure IoT中心和Azure智能云的网络能力，让"听听巴士"技术团队以轻松、可靠和安全的方式，连接百万级别的物联网终端。

计算未来

总结

在更广泛的数字化转型空间里，微软人工智能正在广泛触发以人工智能为抓手、以大数据为基础的新一轮数字化转型浪潮。人工智能、虚拟现实、混合现实、大数据、云计算等新技术是一个完整的体系，它们互相配合形成端到端的用户体验和解决方案，唯有系统和全面地推进所有的新技术，才能真正建立用户对于人工智能的最终信任、打开全新的数字经济市场空间，从而让中国得以参与全球化数字经济发展，贡献中国智慧，承担中国责任。

预见未来的最佳方法就是创造未来。我们相信，人工智能将为人类创造一个更加美好的未来。微软对这样的未来有着充分的理解与信心：今天微软和微软亚太研发集团为中国输送了大量的新技术人才，通过微软人工智能技术与智能云平台促进了中国本地经济的发展和全球化拓展，表达了微软愿意与中国共同讨论新技术所带来的挑战、建立人们对新技术的信任，微软愿意与中国一起承担人工智能的中国使命。

注释

1. Brad Smith，Carol Ann Browner，《科技史上的今天：马儿失业日》；https://www.linkedin.com/pulse/today-technology-day-horse-lost-its-job-brad-smith/。

2. Lendol Calder，《筹资美国梦：消费信贷的文化史》，普林斯顿大学出版社1999年版，第184页。

3. John Steele Gordon，《财富帝国：美国经济力量的史诗》，哈珀柯林斯出版集团2004年版，第299—300页。

4. 沈向洋博客，2017年7月，https://blogs.microsoft.com/blog/2017/07/12/microsofts-role-intersection-ai-people-society。

5. https://blogs.microsoft.com/ai/microsoft-researchers-win-imagenet-computer-vision-challenge.

6. https://www.microsoft.com/en-us/research/blog/microsoft-researchers-achieve-new-conversational-speech-recognition-milestone.

7. 参见沈向洋博客，2017年5月，https://blogs.microsoft.com/blog/2017/05/10/microsoft-build-2017-microsoft-ai-amplify-human-ingenuity。

8. https://www.microsoft.com/en-us/research/project/medical-image-analysis.

9. https://www.microsoft.com/en-us/research/project/project-premonition.

10. 例如，如果你问小娜"爱尔兰有多大？"她不仅仅会回答你爱尔兰的平方公里面积，而且还会说"大约和南加州一样大"。

11. https://www.microsoft.com/en-us/seeing-ai.

12. https://www.microsoft.com/en-us/research/project/farmbeats-iot-agriculture/#.

13. https://www.partnershiponai.org.

14. https://www.nytimes.com/2017/10/26/opinion/algorithm-compas-sentencing-bias.html和https://www.propublica.org/article/machine-bias-risk-assessments-in-criminal-sentencing.

15. https://www.nytimes.com/2017/11/21/magazine/can-ai-be-taught-to-explain-itself.html.

16. Daniel Solove,《信息隐私权法简史》（GW Law）2006年, 第1—25页。

17. 马车到汽车的演变引发了一系列深刻的洞见。这一演变创造出了众多的新兴产业, 其中许多产业在汽车投入使用之初根本无法预见。https://www.linkedin.com/pulse/today-technology-day-horse-lost-its-job-brad-smit.

18. http://www3.weforum.org/docs/WEF_FOJ_Executive_Summary_Jobs.pdf.

19. http://query.nytimes.com/gst/abstract.html?res=9C03EEDF1F39E133A25755C2A9649C946995D6CF&legacy=true.

20. https://www.economist.com/news/special-report/21700758-will-smarter-machines-cause-mass-unemploy-

ment-automation-and-anxiety.

21. https://www.economist.com/news/special-report/21700758-will-smarter-machines-cause-mass-unemployment-auto-mation-and-anxiety.

22. https://www.economist.com/news/special-report/21700758-will-smarter-machines-cause-mass-unemploy-ment-automation-and-anxiety.

23. https://venturebeat.com/2017/10/04/the-fundamental-differences-between-automation-and-ai.

24. https://www.washingtonpost.com/news/theworldpost/wp/2017/10/19/inside-chinas-quest-to-become-the-global-leader-in-ai/?utm_term=.9da300d7d549.

25. 人工智能调查。风险驱动因素。https://news.microsoft.com/cloudforgood/policy/briefing-papers/responsible-cloud/amplifying-human-ingenuity-artificial-intelligence.html.

26. https://www.oxfordmartin.ox.ac.uk/downloads/academic/The_Future_of_Employment.pdf.

27. https://openknowledge.worldbank.org/handle/10986/23347.

28. https://papers.ssrn.com/sol3/papers.cfm?abstract_id=2940245.

29. https://www.theguardian.com/technology/2017/

jan/11/robots-jobs-employees-artificial-intelligence.

30. https://www.postandcourier.com/business/as-amazon-pushes-forward-with-robots-workers-find-new-roles/article_c5777048-97ca-11e7-955e-8f628022e7cc.html.

31. https://www.forrester.com/report/The+Future+Of+Jobs+2025+Working+Side+By+Side+With+Robots/-/E-RES119861.

32. https://www.economist.com/news/special-report/21700758-will-smarter-machines-cause-mass-unemployment-automation-and-anxiety.

33. 《新新工作方法系列：将根本改变组织机构工作方式的十二种力量》，BCG，2017年3月。https://www.bcg.com/en-us/publications/2017/people-organization-strategy-twelve-forces-radically-change-organizations-work.aspx.

34. http://reports.weforum.org/future-of-jobs-2016/skills-stability/?doing_wp_cron=1514488681.1306788921356201171875.

35. https://www.technologyreview.com/s/515926/how-technology-is-destroying-jobs.

36. https://cew.georgetown.edu/wp-content/uploads/Americas-Divided-Recovery-web.pdf.

37. https://krueger.princeton.edu/sites/default/files/akrueger/files/katz_krueger_cws_-_march_29_20165.pdf.

38. http://www.oxfordmartin.ox.ac.uk/publications/view/1314.

39. http://www.hamiltonproject.org/papers/who_is_out_of_the_labor_force.

40. http://www.pewinternet.org/2016/11/17/gig-work-online-selling-and-home-sharing.

41. 根据劳工统计局提供的数字，出于个人意愿选择兼职工作的人员数量达600万，这一数字与2007年相比增加了12%。http://www.bloomberg.com/news/articles/2015-08-18/why-6-million-americans-would-rather-work-part-time.

42. https://www.teacherspayteachers.com.

43. http://journals.sagepub.com/eprint/3FMTvCNPJ4SkhW9tgpWP/full.

44. http://globalworkplaceanalytics.com/resources/costs-benefits.

45. http://www.pewsocialtrends.org/2016/10/06/4-skills-and-training-needed-to-compete-in-todays-economy.

46. 此外，根据国家教育统计中心提供的数据，每五名高中生中，有一名高中生未能在四年内毕业。

47. https://secure-media.collegeboard.org/digitalServices/pdf/research/2016/Program-Summary-Report-2016.pdf.

48. https://www.bls.gov/charts/job-openings-and-labor-turnover/opening-hire-seps-rates.htm.

49. https://www.bloomberg.com/news/articles/2017-06-22/the-worlds-workers-have-bigger-problems-than-a-robot-apocalypse.

50. https://www.nationalskillscoalition.org/resources/publications/2017-middle-skills-fact-sheets/file/United-States-MiddleSkills.pdf.

51. http://burning-glass.com/wp-content/uploads/2015/06/Digital_Skills_Gap.pdf.

52. https://www.nationalskillscoalition.org/resources/publications/file/Opportunity-Knocks-How-expanding-the-Work-Opportunity-Tax-Credit-could-grow-businesses-help-low-skill-workers-and-close-the-skills-gap.pdf.

53. 为偏远和贫困社区提供宽带有助于扩大获得教育、培训的机会，提高教育、培训质量，加强公民参与。但是，目前乡村地区尚有2340万人口无法获得宽带接入，因此无法使用按需学习工具。为了满足这一需求，2017年7月，微软发起了一项"乡村通信计划"，希望推动这一新模式的广泛应用，并助力在2022年7月4日前消除美国乡村的宽带空白。https://news.microsoft.com/rural-broadband.

54. 例如，在微软全球技能计划中，有一个微软印度Oorja项目，这个项目与理工、工业技术研究院和工程学校共同合作，帮助学生们获得各种微软教育课程证书（课程内容主要关乎办公效率），使学生们为就业做好准备。https://www.microsoft.com/en-in/about/citizenship/youthspark/youthsparkhub/programs/partners-in-learning.

55. https://news.microsoft.com/download/presskits/education/docs/IDC_101513.pdf.

56. https://news.microsoft.com/2017/06/27/the-markle-founda-tion-and-microsoft-partner-to-accelerate-a-skills-based-labor-market-for-the-digital-economy.

57. 我们需要更准确和更新的数据来了解不断变化的工作内容及所需的技能，同样，我们也需要更好地了解雇主和雇员关系以及工作条件的变化趋势，包括工作性质的变化。此外，许多现有政府项目通过工资数据来评估就业成果；人们需要更广泛的数据，才能了解新兴的权变劳动安排的真正影响。平台公司可提供私营领域的数据，加强此项分析。

58. 尽管根据大多数的预测，在线平台劳动力人数占比仍不到百分之一，但是处于非传统雇主/雇员劳动关系中的劳动力比例（临时代理人、随叫随到劳动人员、合同工、独立承揽人、自由职业人）却远远高出这一数字。参见《1995—2015年美国替代劳动安排的崛起及性质》。

59. 在立法还没有完成现代化的情况下，监管机构作出了许多法律解释，而这些解释在很大程度上已经偏离早期先例，例如，扩展了联合雇佣的范围。随着许多监管机构政治组成的改变，新的判例法呈现出向相反方面摇摆的趋势。美国议会也提出了对关键定义进行立法的建议。

60. http://www.goodworkcode.org/about.

61. http://www.care.com.

62. 参见参议员Warner和众议员DelBene提出的《独立劳动者移动福利试点项目法案》S.1251和H.R.2685部分。该法案将在美国劳动管理部门设立一个移动福利试点项目，向州、地方政府和非营利机构提供2000万美元的竞争性资助，用以开展新模式的试点和评估工作或对现有模式作出改进，为独立承揽人、临时劳动人员和自雇型劳动人员提供移动福利。

63. 我们通过现有数据了解到，近几十年来，劳动者在职业生涯中曾经历多次失业。国家青年纵向调查1979（NLSY79）从1957年至1964年间的出生者中选取出具有全国代表性的样本，并进行了跟踪；调查显示，这些人在18岁至48岁之间平均经历了5.6次失业。高中辍学的劳动者在18至岁48岁之间平均经历了7.7次失业，而高中毕业和大学毕业的劳动者则分别经历了5.4和3.9次失业波。此外，在调查中，将近三分之一高中辍学的劳动者经历的失业达10次或以上，高中毕业和大学毕业的劳动者的这一比例则分别为22%和6%。

Foreword

The Future Computed

Twenty years ago, we both worked at Microsoft, but on opposite sides of the globe. In 1998, one of us was living and working in China as a founding member of the Microsoft Research Asia lab in Beijing. Five thousand miles away, the other was based at the company's headquarters, just outside of Seattle, leading the international legal and corporate affairs team. While we lived on separate continents and in quite different cultures, we shared a common workplace experience within Microsoft, albeit with differing routines before we arrived at the office.

At that time in the United States, waking to the scent of brewing coffee was a small victory in technology automation. It meant that you had remembered to set the timer on the programmable coffee maker the night before. As you drank that first cup of coffee, you typically watched the morning news on a standard television or turned the pages of the local newspaper to learn what had happened while you slept. For many people a daily diary was your lifeline, reminding you

The Future Computed

of the coming day's activities: a morning meeting at the office, dial-in numbers and passcodes for conference calls, the address for your afternoon doctor's appointment, and a list of to-dos including programming the VCR to record your favorite show. Before you left for the day, you might have placed a few phone calls (and often left messages on answering machines), including to remind sitters when to pick up children or confirm dinner plans.

Twenty years ago, for most people in China, an LED alarm clock was probably the sole digital device in your bedroom. A bound personal calendar helped you track the day's appointments, addresses, and phone numbers. After sending your kids off to school, you likely caught up on the world's happenings from a radio broadcast while you ate a quick breakfast of soya milk with Youtiao at your neighborhood restaurant. In 1998, commuters in Beijing buried their noses in newspapers and books – not smartphones and laptops – on the crowded trains and buses traveling to and from the city's centers.

But today, while many of our fundamental morning routines remain the same, a lot has also changed as technology has altered how we go about them. Today a morning in Beijing is still different from a morning in Seattle, but not as different as it used to be. Consider for a moment that in both places the smartphone charging on your bedside table is the device that not only wakes you, but serves up headlines and updates you on your friends' social lives. You read all the email that arrived overnight, text your sister to confirm dinner plans, update the calendar invite to your sitter with details for soccer practice, and then check traffic conditions. Today, in 2018, you can order and pay for a double skinny latte or tea from Starbucks and request a ride-share to drive you to work from that same smartphone.

Foreword
The Future Computed

Compared with the world just 20 years ago, we take a lot of things for granted that used to be the stuff of science fiction. Clearly much can change in just two decades.

Twenty years from now, what will your morning look like? At Microsoft, we imagine a world where your personal digital assistant Cortana talks with your calendar while you sleep. She works with your other smart devices at home to rouse you at the end of a sleep cycle when it's easiest to wake and ensures that you have plenty of time to shower, dress, commute and prepare for your first meeting. As you get ready, Cortana reads the latest news, research reports and social media activity based on your current work, interests and tasks, all of which she gleaned from your calendar, meetings, communications, projects and writings. She updates you on the weather, upcoming meetings, the people you will see, and when you should leave home based on traffic projections.

Acting on the request you made a year before, Cortana also knows that it's your sister's birthday and she's ordered flowers (lilies, your sister's favorite) to be delivered later that day. (Cortana also reminds you about this so that you'll know to say, "you're welcome" when your sister thanks you.) Cortana has also booked a reservation for a restaurant that you both like at a time that's convenient for both of your schedules.

In 2038, digital devices will help us do more with one of our most precious commodities: time.

In 20 years, you might take your first meeting from home by slipping on a HoloLens or other device where you'll meet and interact with your colleagues and clients around a virtual boardroom powered by mixed reality. Your presentation and remarks will be translated automatically into each participant's native language, which they will hear through an earpiece or phone. A digital assistant like Cortana will then automatically prepare a summary of the meeting with tasks assigned to the participants and reminders placed on their schedules based on the conversation that took place and the decisions the participants made.

In 2038, a driverless vehicle will take you to your first meeting while you finalize a presentation on the car's digital hub. Cortana will summarize research and data pulled from newly published articles and reports, creating infographics with the new information for you to review and accept. Based on your instructions, she'll automatically reply to routine emails and reroute those that can be handled by others, which she will request with a due date based on the project timeline. In fact, some of this is already happening today, but two decades from now everyone will take these kinds of capabilities for granted.

Increasingly, we imagine that a smart device will monitor your health vitals. When something is amiss, Cortana will schedule an appointment, and she will also track and schedule routine checkups, vaccines and tests. Your digital assistant will book appointments and reserve time on your calendar on days that are most convenient. After work a self-driving car will take you home, where you'll join your doctor for a virtual checkup. Your mobile device will take your blood pressure, analyze your blood and oxygen level, and send the results to your doctor, who will analyze the data during your call. Artificial intelligence will help your doctor analyze your results using more than a terabyte of health data, helping her accurately diagnose and prescribe a customized treatment based on your unique physiological traits. Within a few hours, your medication will arrive at your door by drone, which Cortana will remind you to take. Cortana will also monitor your progress and, if you don't improve, she'll ask your permission to book a follow-up appointment with the doctor.

When it's time to take a break from the automated world of the future, you won't call a travel agent or even book online your own flight or hotel as you do today. You'll simply say, "Hey, Cortana, please plan a two-week holiday." She'll propose a custom itinerary based on the season, your budget, availability and interests. You'll then decide where you want to go and stay.

Looking back, it's fascinating to see how technology has transformed the way we live and work over the span of twenty years. Digital technology powered by the cloud has made us smarter and helped us optimize our time, be more productive and communicate with one another more effectively. And this is just the beginning.

Before long, many mundane and repetitive tasks will be handled automatically by AI, freeing us to devote our time and energy to more productive and creative endeavors. More broadly, AI will enable humans to harness vast amounts of data and make breakthrough advances in areas like healthcare, agriculture, education and transportation. We're already seeing how AI-bolstered computing can help doctors reduce medical mistakes, farmers improve yields, teachers customize instruction and researchers unlock solutions to protect our planet.

But as we've seen over the past 20 years, as digital advances bring us daily benefits they also raise a host of complex questions and broad concerns about how technology will affect society. We have seen this as the internet has come of age and become an essential part of our work and private lives. The impact ranges from debates around the dinner table about how distracting our smartphones have become to public deliberations about cybersecurity, privacy, and even the role social media plays in terrorism. This has given birth not just to new public policies and regulations, but to new fields of law and to new ethical considerations in the field of computer science. And this seems certain to continue as AI evolves and the world focuses on the role it will play in society. As we look to the future, it's important that we maintain an open and questioning mind while we seek to take advantage of the opportunities and address the challenges that this new technology creates.

The development of privacy rules over the past two decades provides a good preview of what we might expect to see more broadly in the coming years for issues relating to AI. In 1998, one would have been hard-pressed to find a full-time "privacy lawyer." This legal discipline was just emerging with the advent of the initial digital privacy laws, perhaps most notably the European Community's Data Protection Directive, adopted in 1995. But the founding of the International Association of Privacy Professionals, or IAPP, the leading professional organization in the field, was still two years away.

Today, the IAPP has over 20,000 members in 83 countries. Its meetings take place in large convention centers filled with thousands of people. There's no shortage of topics for IAPP members to discuss, including questions of corporate responsibility and even ethics when it comes to the collection, use, and protection of consumer information. There's also no lack of work for privacy lawyers now that data protection agencies -the privacy regulators of our age -are operating in over 100 countries. Privacy regulation, a branch of law that barely existed two decades ago, has become one of the defining legal fields of our time.

What will the future bring when it comes to the issues, policies and regulations for artificial intelligence? In computer science, will concerns about the impact of AI mean that the study of ethics will become a requirement for computer programmers and researchers? We believe that's a safe bet. Could we see a Hippocratic Oath for coders like we have for doctors? That could make sense. We'll all need to learn together and with a strong commitment to broad societal responsibility. Ultimately the question is not only what computers can do. It's what computers should do.

Similarly, will the future give birth to a new legal field called "AI law"? Today AI law feels a lot like privacy law did in 1998. Some existing laws already apply to AI, especially tort and privacy law, and we're starting to see a few specific new regulations emerge, such as for driverless cars. But AI law doesn't exist as a distinct field. And we're not yet walking into conferences and meeting people who introduce themselves as "AI lawyers." By 2038, it's safe to assume that the situation will be different. Not only will there be AI lawyers practicing AI law, but these lawyers, and virtually all others, will rely on AI itself to assist them with their practice.

The real question is not whether AI law will emerge, but how it can best come together -and over what timeframe. We don't have all the answers, but we're fortunate to work every day with people who are asking the right questions. As they point out, AI technology needs to continue to develop and mature before rules can be crafted to govern it. A consensus then needs to be reached about societal principles and values to govern AI development and use, followed by best practices to live up to them. Then we're likely to be in a better position for governments to create legal and regulatory rules for everyone to follow.

This will take time -more than a couple of years in all likelihood, but almost certainly less than two decades. Already it's possible to start defining six ethical principles that should guide the development and use of artificial intelligence. These principles should ensure that AI systems are fair, reliable and safe, private and secure, inclusive, transparent, and accountable. The more we build a detailed understanding of these or similar principles -and the more technology developers and users can share best practices to implement them -the better served the world will be as we begin to contemplate societal rules to govern AI.

Today, there are some people who might say that ethical principles and best practices are all that is needed as we move forward. They suggest that technology innovation doesn't really need the help of regulators, legislators and lawyers.

While they make some important points, we believe this view is unrealistic and even misguided. AI will be like every technology that has preceded it. It will confer enormous benefits on society. But inevitably, some people will use it to cause harm. Just as the advent of the postal service led criminals to invent mail fraud and the telegraph was followed by wire fraud, the years since 1998 have seen both the adoption of the internet as a tool for progress and the rise of the internet as a new arena for fraud, practiced in increasingly creative and disturbing ways on a global basis.

We must assume that by 2038, we'll grapple with the issues that arise when criminal enterprises and others use AI in ways that are objectionable and even harmful. And undoubtedly other important questions will need to be addressed regarding societally acceptable uses for AI. It will be impossible to address these issues effectively without a new generation of laws. So, while we can't afford to stifle AI technology by adopting laws before we understand the issues that lie ahead of us, neither can we make the mistake of doing nothing now and waiting for two decades before getting started. We need to strike a balance.

As we consider principles, policies and laws to govern AI, we must also pay attention to AI's impact on workers around the globe. What jobs will AI eliminate? What jobs will it create? If there has been one constant over 250 years of technological change, it has been the ongoing impact of technology on jobs -the creation of new jobs, the elimination of existing jobs, and the evolution of job tasks and content. This too is certain to continue with the adoption of AI.

Will AI create more jobs than it will eliminate? Or will it be the other way around? Economic historians have pointed out that each prior industrial revolution created jobs on a net basis. There are many reasons to think this will also be the case with AI, but the truth is that no one has a crystal ball.

It's difficult to predict detailed employment trends with certainty because the impact of new technology on jobs is often indirect and subject to a wide range of interconnected innovations and events. Consider the automobile. One didn't need to be a soothsayer to predict that the adoption of cars would mean fewer jobs producing horse-drawn carriages and new jobs manufacturing automobile tires. But that was just part of the story.[1]

The transition to cars initially contributed to an agricultural depression that affected the entire American economy in the 1920s and 1930s. Why? Because as the horse population declined rapidly, so did the fortunes of American farmers. In the preceding decade roughly a quarter of agricultural output had been used to feed horses. But fewer horses meant less demand for hay, so farmers shifted to other crops, flooding the market and depressing agricultural prices more broadly. This agricultural depression impacted local banks in rural areas, and then this rippled across the entire financial system.

Other indirect effects had a positive economic impact as the sale of automobiles led to the expansion of industry sectors that, at first glance, appear disconnected from cars. One example was a new industry to provide consumer credit. Henry Ford's invention of the assembly line made cars affordable to a great many families, but cars were still expensive and people needed to borrow money to pay for them. As one historian noted, "installment credit and the automobile were both cause and consequence of each other's success."[2] In short, a new financial services market took flight.

Something similar happened with advertising. As passengers traveled in cars driving 30 miles per hour or more, "a sign had to be grasped instantly or it wouldn't be grasped at all."[3] Among other things, this led to the creation of corporate logos that could be recognized immediately wherever they appeared.

Consider the indirect impact of the automobile on the island of Manhattan alone. The cars driving down Broadway contributed to the creation of new financial jobs on Wall Street and new advertising positions on Madison Avenue. Yet there's little indication that anyone predicted either of these new job categories when cars first appeared on city streets.

One of the lessons for AI and the future is that we'll all need to be alert and agile to the impact of this new technology on jobs. While we can predict generally that new jobs will be created and some existing jobs will disappear, none of us should develop such a strong sense of certainty that we lose the ability to adapt to the surprises that probably await us.

But as we brace ourselves for uncertainty, one thing remains clear. New jobs will require new skills. Indeed, many existing jobs will also require new skills. That is what always happens in the face of technological change.

Consider what we've seen over the past three decades. Today every organization of more than modest size has one or more employees who support its IT, or information technology. Very few of these jobs existed 30 years ago. But it's not just IT staff that needed to acquire IT skills. In the early 1980s, people in offices wrote with a pen on paper, and then secretaries used typewriters to turn that prose into something that was actually legible. By the end of the decade, secretaries learned to use word processing terminals. And then in the 1990s, everyone learned to do their own writing on a PC and the number of secretaries declined. IT training wasn't just reserved for IT professionals.

In a similar way, we're already seeing increasing demand for new digital and other technical skills, with critical shortages appearing in some disciplines. This is expanding beyond coding and computer science to data science and other fields that are growing in importance as we enter the world's Fourth Industrial Revolution. More and more, this isn't just a question of encouraging people to learn new skills, but of finding new ways to help them acquire the skills they will need. Surveys of parents show that they overwhelmingly want their children to have the opportunity to learn to code. And at Microsoft, when we offer our employees new courses on the latest AI advances, demand is always extremely high.

The biggest challenges involve the creation of ways to help people learn new skills, and then rethinking how the labor market operates to enable employers and employees to move in more agile ways to fill new positions. The good news is that many communities and countries have developed new innovations to address this issue, and there are opportunities to learn from these emerging practices. Some are new approaches to longstanding programs, like Switzerland's successful youth apprenticeships. Others are more recent innovations spurred by entities such as LinkedIn and its online tools and services and nonprofit ventures like the Markle Foundation's Skillful initiative in Colorado.

The impact of AI, the cloud and other new technologies won't stop there. A few decades ago, workers in many countries mostly enjoyed traditional employer-employee relationships and worked in offices or manufacturing facilities. Technology has helped upend this model as more workers engage in alternative work arrangements through remote and part-time work, as contractors or through project-based engagements. And most studies suggest that these trends will continue.

For AI and other technologies to benefit people as broadly as possible, we'll need to adapt employment laws and labor policies to address these new realities. Many of our current labor laws were adopted in response to the innovations of the early 20th century. Now, a century later, they're no longer suited to the needs of either workers or employers. For example, employment laws in most countries assume that everyone is either a full-time employee or an independent contractor, making no room for people who work in the new economy for Uber, Lyft or other similar services that are emerging in every field from tech support to caregiving.

Similarly, health insurance and other benefits were designed for full-time employees who remain with a single employer for many years. But they aren't as effective for individuals who work for multiple companies simultaneously or change jobs more frequently. Our social safety net -including the United States' Social Security system -is a product of the first half of the last century. There is an increasingly pressing need to adapt these vital public policies to the world that is changing today.

As we all think about the future, the pace of change can feel more than a little daunting. By looking back to technology in 1998, we can readily appreciate how much change we've lived through already. Looking ahead to 2038, we can begin to anticipate the rapid changes that lie ahead -changes that will create opportunities and challenges for communities and countries around the world.

For us, some key conclusions emerge

First, the companies and countries that will fare best in the AI era will be those that embrace these changes rapidly and effectively. The reason is straightforward: AI will be useful wherever intelligence is useful, helping us to be more productive in nearly every field of human endeavor and leading to economic growth. Put simply, new jobs and economic growth will accrue to those that embrace the technology, not those that resist it.

Second, while we believe that AI will help improve daily life in many ways and help solve big societal problems, we can't afford to look to this future with uncritical eyes. There will be challenges as well as opportunities. This is why we need to think beyond the technology itself to address the need for strong ethical principles, the evolution of laws, the importance of training for new skills, and even labor market reforms. This must all come together if we're going to make the most of this new technology.

Third, we need to address these issues together with a sense of shared responsibility. In part this is because AI technology won't be created by the tech sector alone. At Microsoft we're working to "democratize AI" in a manner that's similar to the way we "democratized the PC." Just as our work that started in the 1970s enabled organizations across society to create their own custom applications for the PC, the same thing will happen with AI. Our approach to AI is making the fundamental AI building blocks like computer vision, speech, and knowledge recognition available to every individual and organization to build their own AI-based solutions. We believe this is far preferable to having only a few companies control the future of AI. But just as this will spread broadly the opportunity for others to create AI-based systems, it will spread broadly the shared responsibility needed to address AI issues and their implications.

As technology evolves so quickly, those of us who create AI, cloud and other innovations will know more than anyone else how these technologies work. But that doesn't necessarily mean that we will know how best to address the role they should play in society. This requires that people in government, academia, business, civil society, and other interested stakeholders come together to help shape this future. And increasingly we need to do this not just in a single community or country, but on a global basis. Each of us has a responsibility to participate -and an important role to play.

All of this leads us to what may be one of the most important conclusions of all. We're reminded of something that Steve Jobs famously talked about repeatedly: he always sought to work at the intersection of engineering and the liberal arts.

One of us grew up learning computer science and the other started in the liberal arts. Having worked together for many years at Microsoft, it's clear to both of us that it will be even more important to connect these fields in the future.

At one level, AI will require that even more people specialize in digital skills and data science. But skilling-up for an AI-powered world involves more than science, technology, engineering and math. As computers behave more like humans, the social sciences and humanities will become even more important. Languages, art, history, economics, ethics, philosophy, psychology and human development courses can teach critical, philosophical and ethics-based skills that will be instrumental in the development and management of AI solutions. If AI is to reach its potential in serving humans, then every engineer will need to learn more about the liberal arts and every liberal arts major will need to learn more about engineering.

We're all going to need to spend more time talking with, listening to, and learning from each other. As two people from different disciplines who've benefited from doing just that, we appreciate firsthand the valuable and even enjoyable opportunities this can create.

We hope that the pages that follow can help as we all get started.

Acknowledgements

We would like to thank the following contributors for providing their insights and perspectives in the development of this book.

Benedikt Abendroth, Geff Brown, Carol Ann Browne, Dominic Carr, Pablo Chavez, Steve Clayton, Amy Colando, Jane Broom Davidson, Mariko Davidson, Paul Estes, John Galligan, Sue Glueck, Cristin Goodwin, Mary Gray, David Heiner, Merisa Heu-Weller, Eric Horvitz, Teresa Hutson, Nicole Isaac, Lucas Joppa, Aaron Kleiner, Allyson Knox, Cornelia Kutterer, Jenny Lay-Flurrie, Andrew Marshall, Anne Nergaard, Carolyn Nguyen, Barbara Olagaray, Michael Philips, Brent Sanders, Mary Snapp, Dev Stahlkopf, Steve Sweetman, Lisa Tanzi, Ana White, Joe Whittinghill, Joshua Winter, Portia Wu.

CONTENTS

Chapter 1
The Future of Artificial Intelligence   151

Microsoft's Approach to AI   159
The Potential of Modern AI -   167
Addressing Societal Challenges
The Challenges AI Presents   171

Chapter 2
Principles, Policies and Laws for the   173
Responsible Use of AI

Ethical and Societal Implications   176
Developing Policy and Law for Artificial Intelligence   189
Fostering Dialogue and the Sharing of Best Practices   198

Chapter 3
AI and the Future of Jobs and Work   201

The Impact of Technology on Jobs and Work   205
The Changing Nature of Work, the Workplace and Jobs   213
Preparing Everyone for the Future of Work   219

Changing Norms for Changing Worker Needs	232
Working Together	242

Conclusion
AI Amplifying Human Ingenuity — 243

Addendum
China's Mission in the Future of AI — 249

Digital Economy Brings Global Inclusive Opportunities	254
Innovation in China, Innovation for the Whole World	256
With Great Power Comes Great Responsibility	259
The AI Path with Chinese Characteristics	261
Conclusion	267

Notes — 269

Chapter 1
The Future of Artificial Intelligence

> **I propose to consider the question. Can machines think?**

Alan Turing, 1950

Chapter 1
The Future
of Artificial
Intelligence

*I*n the summer of 1956, a team of researchers at Dartmouth College met to explore the development of computer systems capable of learning from experience, much as people do. But, even this seminal moment in the development of AI was preceded by more than a decade of exploration of the notion of machine intelligence, exemplified by Alan Turing's quintessential test: a machine could be considered "intelligent" if a person interacting with it (by text in those days) could not tell whether it was a human or a computer.

Researchers have been advancing the state of the art in AI in the decades since the Dartmouth conference. Developments in subdisciplines such as machine vision, natural language understanding, reasoning, planning and robotics have produced an ongoing stream of innovations, many of which have already become part of our daily lives. Route-planning features in navigation systems, search engines that retrieve and rank content from the vast amounts of information on the internet, and machine vision capabilities that enable postal services to automatically recognize and route handwritten addresses are all enabled by AI.

The Future Computed

At Microsoft, we think of AI as a set of technologies that enable computers to perceive, learn, reason and assist in decision-making to solve problems in ways that are similar to what people do. With these capabilities, how computers understand and interact with the world is beginning to feel far more natural and responsive than in the past, when computers could only follow pre-programmed routines.

Not so long ago we interacted with computers via a command line interface. And while the graphical user interface was an important step forward, we will soon be routinely interacting with computers just by talking to them, just as we would to a person. To enable these new capabilities, we are, in effect, teaching computers to see, hear, understand and reason. [4]Key technologies include:

Vision: the ability of computers to "see" by recognizing what is in a picture or video.

Speech: the ability of computers to "listen" by understanding the words that people say and to transcribe them into text.

Language: the ability of computers to "comprehend" the meaning of the words, taking into account the many nuances and complexities of language (such as slang and idiomatic expressions).

Knowledge: the ability of a computer to "reason" by understanding the relationship between people, things, places, events and the like. For instance, when a search result for a movie provides information about the cast and other movies those actors were in, or at work when you participate in a meeting and the last several documents that you shared with the person you're meeting with are automatically delivered to you. These are examples of a computer reasoning by drawing conclusions about which information is related to other information.

Chapter 1
The Future of Artificial Intelligence

Computers are learning the way people do; namely, through experience. For computers, experience is captured in the form of data. In predicting how bad traffic will be, for example, computers draw upon data regarding historical traffic flows based on the time of day, seasonal variations, the weather, and major events in the area such as concerts or sporting events. More broadly, rich "graphs" of information are foundational to

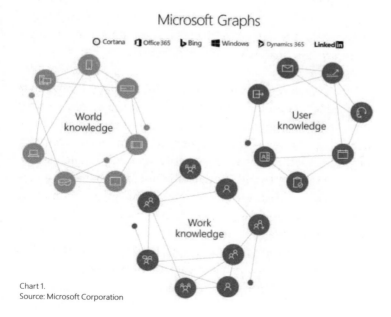

Chart 1.
Source: Microsoft Corporation

enabling computers to develop an understanding of relevant relationships and interactions between people, entities and events. In developing AI systems, Microsoft is drawing upon graphs of information that include knowledge about the world, about work and about people.

Thanks in part to the availability of much more data, researchers have made important strides in these technologies in the past few years. In 2015, researchers at Microsoft announced that they had taught computers to identify objects in a photograph or video as accurately as people do in a test using the standard ImageNet 1K database of images.[5] In 2017, Microsoft's researchers announced they had developed a speech recognition system that understood spoken words as accurately as a team of professional transcribers, with an error rate of just 5.1 percent using the standard Switchboard dataset.[6] In essence, AI-enhanced computers can, in most cases, see and hear as accurately as humans.

Much work remains to be done to make these innovations applicable to everyday use. Computers still may have a hard time understanding speech in a noisy environment where people speak over one another or when presented with unfamiliar accents or languages. It is especially challenging to teach computers to truly understand not just what words were spoken, but what the words mean and to reason by drawing conclusions and making decisions based on them. To enable computers to comprehend meaning and answer more complex questions, we need to take a big-picture view, understand and evaluate context, and bring in background knowledge.

Why Now?

Researchers have been working on AI for decades. Progress has accelerated over the past few years thanks in large part to three developments: the increased availability of data; growing cloud computing power; and more powerful algorithms developed by AI researchers.

As our lives have become increasingly digitized and sensors have become cheap and ubiquitous, more data than ever before is available for computers to learn from.

Chart 2.
Source: IDC Digital Universe Forecast, 2014

Only with data can computers discern the patterns, often subtle, that enable them to "see," "hear" and "understand."

Analyzing all this data requires massive computing power, which is available thanks to the efficiencies of cloud computing. Today, organizations of any type can tap into the power of the cloud to develop and run their AI systems.

The Future Computed

Researchers at Microsoft, other technology firms, universities and governments have drawn upon this combination of the availability of this data, and with it ready access to powerful computing and breakthroughs in AI techniques -such as "deep learning" using so-called "deep neural nets"-to enable computers to mimic how people learn.

In many ways, AI is still maturing as a technology. Most of the progress to date has been in teaching computers to perform narrow tasks -play a game, recognize an image, predict traffic. We have a long way to go to imbue computers with "general" intelligence. Today's AI cannot yet begin to compete with a child's ability to understand and interact with the world using senses such as touch, sight and smell. And AI systems have only the most rudimentary ability to understand human expression, tone, emotion and the subtleties of human interaction. In other words, AI today is strong on "IQ" but weak on "EQ."

At Microsoft, we're working toward endowing computers with more nuanced capabilities. We believe an integrated approach that combines various AI disciplines will lead to the development of more sophisticated tools that can help people perform more complex, multifaceted tasks. Then, as we learn how to combine multiple IQ functions with abilities that come naturally to people -like applying knowledge of one task to another, having a commonsense understanding of the world, interacting naturally, or knowing when someone is trying to be funny or sarcastic, and the difference between those -AI will become even more helpful. While this is clearly a formidable challenge, when machines can integrate the smarts of IQ and the empathy of EQ in their interactions, we will have achieved what we call "conversational AI." This will be an important step forward in the evolution of computer-human interaction.

Chapter 1
The Future of Artificial Intelligence

Microsoft's Approach to AI

When Bill Gates and Paul Allen founded Microsoft over 40 years ago, their aim was to bring the benefits of computing -then largely locked up in mainframes -to everyone. They set out to build a "personal" computer that would help people be more productive at home, at school and at work. Today, Microsoft is aiming to do much the same with AI. We're building AI systems that are designed to amplify natural human ingenuity. We're deploying AI systems with the goal of making them available to everyone and aspiring to build AI systems that reflect timeless societal values so that AI earns the trust of all.[7]

Amplifying Human Ingenuity

We believe that AI offers incredible opportunities to drive widespread economic and social progress. The key to attaining these benefits is to develop AI in such a way that it is human-centered. Put simply, we aim to develop AI in order to augment human abilities, especially humankind's innate ingenuity. We want to combine the capabilities of computers with human capabilities to enable people to achieve more.

Computers are very good at remembering things. Absent a system failure, computers never forget. Computers are very good at probabilistic reasoning, something many people are not so good at. Computers are very good at discerning patterns in data that are too subtle for people to notice. With these capabilities, computers can help us make better decisions. And this is a real benefit, because, as researchers in cognitive psychology have established, human decision-making is often imperfect. Broadly speaking, the kind of "computational intelligence" that computers can provide will have a significant impact in almost any field where intelligence itself has a role to play.

•**AI improving medical image analysis for clinicians**

AI systems are already helping people tackle big problems. A good example of this is "InnerEye," a project in which U.K.-based researchers at Microsoft have teamed up with oncologists to develop an AI system to help treat cancer more effectively. [8]

InnerEye uses AI technology originally developed for video gameplay to analyze computed tomography (CT) and magnetic resonance imaging (MRI), and helps oncologists target cancer treatment more quickly. CT and MRI scans allow doctors to look inside a patient's body in three dimensions and study anomalies, such as tumors. For cancer patients who are undergoing radiation therapy, oncologists use such scans to delineate tumors from the surrounding healthy tissue, bone and organs. In turn, this helps focus the cell-damaging radiation treatment on the tumor while avoiding healthy anatomy as much as possible. Today, this 3-D delineation task is manual, slow and error-prone. It requires a radiation oncologist to draw contours on hundreds of cross-sectional images by hand, one at a time -a process that can take hours. InnerEye is being designed to accomplish the same task in a fraction of that time, while giving oncologists full control over the accuracy of the final delineation.

To create InnerEye's automatic segmentation, researchers used hundreds of raw CT and MRI scans (with all identifying patient information removed). The scans were fed into an AI system that learned to recognize tumors and healthy anatomical structures with a clinical level of accuracy. As part of the process, once the InnerEye automatic segmentation is complete, the oncologist goes in to fine-tune the contours. The doctor is in control at all times. With further advances, InnerEye may be helpful for measuring and tracking tumor changes over time, and even assessing whether a treatment is working.

• **AI helping researchers prevent disease outbreaks**

Another interesting example is "Project Premonition." We've all seen the heartbreaking stories of lives lost in recent years to dangerous diseases like Zika, Ebola and dengue that are transmitted from animals and insects to people. Today, epidemiologists often don't learn about the emergence of these pathogens until an outbreak is underway. But this project -developed by scientists and engineers at Microsoft Research, the University of Pittsburgh, the University of California Riverside and Vanderbilt University -is exploring ways to detect pathogens in the environment so public health officials can protect people from transmission before an outbreak begins.[9]

What epidemiologists need are sensors that can detect when pathogens are present. The researchers on this project hit upon an ingenious idea: why not use mosquitoes as sensors? There are plenty of them and they feed on a wide range of animals, extracting a small amount of blood that contains genetic information about the animal bitten and pathogens circulating in the environment.

The researchers use advanced autonomous drones capable of navigating through complex environments to identify areas where mosquitoes breed. They then deploy robotic traps that can distinguish between the types of mosquitoes researchers want to collect and other insects, based on wing movement patterns. Once specimens are collected, cloud-scale genomics and advanced AI systems identify the animals that the mosquitoes have fed on and the pathogens that the animals carry. In the past, this kind of genetic analysis could take a month; now the AI capabilities of Project Premonition have shortened that to about 12 hours.

During a Zika outbreak in 2016, Project Premonition drones and traps were tested in Houston. More than 20,000 mosquitoes were collected from nine different species, including those known to carry Zika, dengue, West Nile virus and malaria. Because the traps also gather data on environmental conditions when an insect is collected, the test provided useful data not only about pathogens in the environment but also about mosquito behavior. This helped Project Premonition researchers improve their ability to target hotspots where mosquitoes breed. Researchers are also working to improve how to identify known diseases and detect the presence of previously unknown pathogens.

While the project is still in its early stages, it may well point the way toward an effective early warning system that will detect some of the world's most dangerous diseases in the environment and help prevent deadly outbreaks.

Making Human-Centered AI Available to All

We cannot deliver on the promise of AI unless we make it broadly available to all. People around the world can benefit from AI -but only if AI technologies are available for them. For Microsoft, this begins with basic R&D. Microsoft Research, with its 26-year history, has established itself as one of the premier research organizations in the world contributing both to the advancement of computer science and to Microsoft products and services. Our researchers have published more than 22,000 papers in all areas of study -from the environment to health, and from privacy to security. Recently, we announced the creation of Microsoft Artificial Intelligence and Research, a new group that brings together approximately 7,500 computer scientists, researchers and engineers. This group is chartered with pursuing a deeper understanding of the computational foundations of intelligence, and is focused on integrating research from all fields of AI research in order to solve some of AI's most difficult challenges.

Chapter 1
The Future of Artificial Intelligence

We continue to encourage researchers to publish their results broadly so that AI researchers around the world -at universities, at other companies and in government settings -can build on these advances.

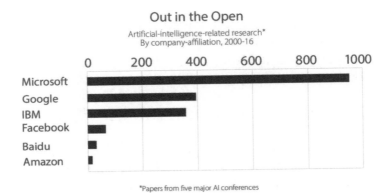

Chart 3.
Source: The Economist

For our customers, we're building AI capabilities into our most popular products, such as Windows and Office. Windows is more secure thanks to AI systems that detect malware and automatically protect computers against it. In Office, Researcher for Word helps you write more compelling documents. Without leaving a document, you can find and incorporate relevant information from across the web using Bing "Knowledge Graph." If you are creating a PowerPoint presentation, PowerPoint Designer assesses the images and text you've used, and provides design tips to create more professional-looking slides, along with suggestions for text captions for images to improve accessibility. And PowerPoint Presentation Translator lets you engage diverse audiences more effectively by breaking down language barriers through auto-captioning in over 60 languages. This feature will also aid people with hearing loss.

AI is the enabling technology behind Cortana, Microsoft's personal digital assistant. Cortana is young, but she's learning fast. Already Cortana can help you schedule a meeting, make a restaurant reservation and find answers to questions on a broad range of topics. Over time, Cortana will be able to interact with other personal digital assistants to automatically handle tasks that take up time and follow familiar patterns. One of the key technologies that Cortana builds upon is Bing, our search service. But instead of just providing links to relevant information, Cortana uses Bing to discover answers to your questions and provide them in a variety of more context-rich ways.[10]

Microsoft is not only using AI technologies to create and enhance our own products, we are also making them available to developers so that they can build their own AI-powered products. The Microsoft AI Platform offers services, tools and infrastructure making AI development easier for developers and organizations of any size. Our service offerings include Microsoft Cognitive Services, a set of pre-built AI capabilities including vision, speech, language and search. All of these are hosted in the cloud and can be easily integrated into applications. Some of these are also customizable so that they can be better optimized to help transform and improve business processes specific to an organization's industry and business needs. You can see the breadth of these offerings below.

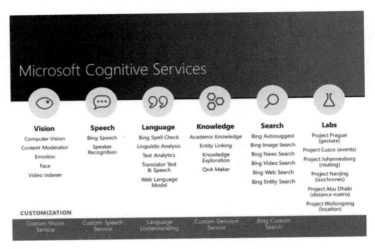

Chart 4.
Source: Microsoft Corporation

We also have technologies available to simplify the creation of "bots" that can engage with people more naturally and conversationally. We offer a growing collection of coding and management tools to make the AI development process easier. And our infrastructure offerings help others develop and deploy algorithms, and store their data and derive insights from it.

Finally, with Microsoft's AI Business Solutions, we are building systems of intelligence so organizations can better understand and act on the information they collect in order to be more productive.

One example of an AI Business Solution is Customer Care Intelligence, currently being used by the Department of Human Services (DHS) in Australia to transform how it delivers services to citizens. At the heart of the program is an expert system that uses a virtual assistant named "Roxy" who helps claims processing officers answer questions and solve problems. Roxy was trained using the DHS operational blueprint that includes all of the agency's policies and procedures, and fed all of the questions that passed between claims officers and DHS managers over a three-month period. In early use, the system was able to answer nearly 80 percent of the questions it was asked. This is expected to translate to about a 20 percent reduction in workload for claims officers.

The internal project with Roxy was so successful that DHS is now developing virtual assistants that will interact directly with citizens. One of these projects will target high school seniors to help them decide whether to apply for a university or enroll in a vocational program through Australia's Technical and Further Education program by helping them navigate the qualification process.

The Potential of Modern AI – Addressing Societal Challenges

At Microsoft, we aim to develop AI systems that will enable people worldwide to more effectively address local and global challenges, and to help drive progress and economic opportunity.

Today's AI enables faster and more profound progress in nearly every field of human endeavor, and it is essential to enabling the digital transformation that is at the heart of worldwide economic development. Every aspect of a business or organization-from engaging with customers to transforming products, optimizing operations and empowering employees -can benefit from this digital transformation.

But even more importantly, AI has the potential to help society overcome some of its most daunting challenges. Think of the most complex and pressing issues that humanity faces: from reducing poverty and improving education, to delivering healthcare and eradicating diseases, addressing sustainability challenges such as growing enough food to feed our fast-growing global population through to advancing inclusion in our society. Then imagine what it would mean in lives saved, suffering alleviated and human potential unleashed if we could harness AI to help us find solutions to these challenges.

Providing effective healthcare at a reasonable cost to the approximately 7.5 billion people on the planet is one of society's most pressing challenges. Whether it's analyzing massive amounts of patient data to uncover hidden patterns that can point the way toward better treatments, identifying compounds that show promise as new drugs or vaccines, or unlocking the potential of personal medicine based on in-depth genetic analysis, AI offers vast opportunities to transform how we understand disease and improve health. Machine reading can help doctors quickly find important information amid thousands of documents that they otherwise wouldn't have time to read. By doing so, it can help medical professionals spend more of their time on higher value and potentially lifesaving work.

Providing safe and efficient transportation is another critical challenge where AI can play an important role. AI-controlled driverless vehicles could reduce traffic accidents and expand the capacity of existing road infrastructure, saving hundreds of thousands of lives every year while improving traffic flow and reducing carbon emissions. These vehicles will also facilitate greater inclusiveness in society by enhancing the independence of those who otherwise are not able to drive themselves.

In education, the ability to analyze how people acquire knowledge and then use that information to develop predictive models for engagement and comprehension points the way toward new approaches to education that combine online and teacher-led instruction and may revolutionize how people learn.

As demonstrated by Australia's Department of Human Services' use of the natural language capabilities of Customer Care Intelligence to answer questions, AI also has the potential to improve how governments interact with their citizens and deliver services.

Chapter 1
The Future of Artificial Intelligence

- **AI enabling people with low vision to hear information about the world around them**

Another area where AI has the potential to have a significant positive impact is in serving the more than 1 billion people in the world with disabilties. One example of how AI can make a difference is a recent Microsoft offering called "Seeing AI," available in the iOS app store, that can assist people with blindness and low vision as they navigate daily life.

Seeing AI was developed by a team that included a Microsoft engineer who lost his sight at 7 years of age. This powerful app, while still in its early stages, demonstrates the potential for AI to empower people with disabilities by capturing images from the user's surroundings and instantly describing what is happening. For example, it can read signs and menus, recognize products through barcodes, interpret handwriting, count currency, describe scenes and objects in the vicinity, or, during a meeting, tell the user that there is a man and a woman sitting across the table who are smiling and paying close attention.[11]

- **AI empowering farmers to be more productive and increase their yield**

And with the world's population expected to grow by nearly 2.5 billion people over the next quarter century, AI offers significant opportunities to increase food production by improving agricultural yield and reducing waste. For example, our "FarmBeats" project uses advanced technology, existing connectivity infrastructure, and the power of the cloud and machine learning to enable data-driven farming at low cost. This initiative provides farmers with easily interpretable insights to help them improve agricultural yield, lower overall costs and reduce the environmental impact of farming.[12]

Given the significant benefits that stem from using AI -empowering us all to accomplish more by being more productive and efficient, driving better business outcomes, delivering more effective government services and helping to solve difficult societal issues-it's vital that everyone has the opportunity to use it. Making AI available to all people and organizations is foundational to enabling everyone to capitalize on the opportunities AI presents and share in the benefits it delivers.

The Challenges AI Presents

As with the great advances of the past on which it builds -including electricity, the telephone and transistors -AI will bring about vast changes, some of which are hard to imagine today. And, as was the case with these previous significant technological advances, we'll need to be thoughtful about how we address the societal issues that these changes bring about. Most importantly, we all need to work together to ensure that AI is developed in a responsible manner so that people will trust it and deploy it broadly, both to increase business and personal productivity and to help solve societal problems.

This will require a shared understanding of the ethical and societal implication of these new technologies. This, in turn, will help pave the way toward a common framework of principles to guide researchers and developers as they deliver a new generation of AI-enabled systems and capabilities, and governments as they consider a new generation of rules and regulations to protect the safety and privacy of citizens and ensure that the benefits of AI are broadly accessible.

In Chapter 2, we offer our initial thinking on how to move forward in a way that respects universal values and addresses the full range of societal issues that AI will raise, while ensuring that we achieve the full potential of AI to create opportunities and improve lives.

Chapter 2
Principles, Policies and Laws for the Responsible Use of AI

> **In a sense, artificial intelligence will be the ultimate tool because it will help us build all possible tools.**

K. Eric Drexler

Chapter 2
Principles, Policies and Laws for the Responsible Use of AI

As AI begins to augment human understanding and decision-making in fields like education, healthcare, transportation, agriculture, energy and manufacturing, it will raise new societal questions. How can we ensure that AI treats everyone fairly? How can we best ensure that AI is safe and reliable? How can we attain the benefits of AI while protecting privacy? How do we not lose control of our machines as they become increasingly intelligent and powerful?

The people who are building AI systems are, of course, required to comply with the broad range of laws around the world that already govern fairness, privacy, injuries resulting from unreasonable behaviors and the like. There are no exceptions to these laws for AI systems. But we still need to develop and adopt clear principles to guide the people building, using and applying AI systems. Industry groups and others should build off these principles to create detailed best practices for key aspects of the development of AI systems, such as the nature of the data used to train AI systems, the analytical techniques deployed, and how the results of AI systems are explained to people using those systems.

It's imperative that we get this right if we're going to prevent mistakes. Otherwise people may not fully trust AI systems. And if people don't trust AI systems, they will be less likely to contribute to the development of such systems and to use them.

Ethical and Societal Implications

Business leaders, policymakers, researchers, academics and representatives of nongovernmental groups must work together to ensure that AI-based technologies are designed and deployed in a manner that will earn the trust of the people who use them and the individuals whose data is being collected. The Partnership on AI (PAI), an organization co-founded by Microsoft, is one vehicle for advancing these discussions. Important work is also underway at many universities and governmental and non-governmental organizations.[13]

Designing AI to be trustworthy requires creating solutions that reflect ethical principles that are deeply rooted in important and timeless values. As we've thought about it, we've focused on six principles that we believe should guide the development of AI. Specifically, AI systems should be fair, reliable and safe, private and secure, inclusive, transparent, and accountable. These principles are critical to addressing the societal impacts of AI and building trust as the technology becomes more and more a part of the products and services that people use at work and at home every day.

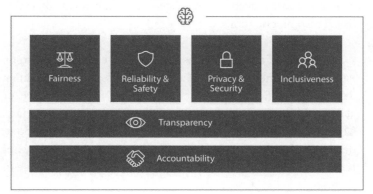

Chart 5.
Source: Microsoft Corporation

Chapter 2
Principles, Policies and Laws for the Responsible Use of AI

Fairness – AI systems should treat all people fairly

AI systems should treat everyone in a fair and balanced manner and not affect similarly situated groups of people in different ways. For example, when AI systems provide guidance on medical treatment, loan applications or employment, they should make the same recommendations for everyone with similar symptoms, financial circumstances or professional qualifications. If designed properly, AI can help make decisions that are fairer because computers are purely logical and, in theory, are not subject to the conscious and unconscious biases that inevitably influence human decision-making. Yet, because AI systems are designed by human beings and the systems are trained using data that reflects the imperfect world in which we live, AI can operate unfairly without careful planning. To ensure that fairness is the foundation for solutions using this new technology, it's imperative that developers understand how bias can be introduced into AI systems and how it can affect AI-based recommendations.

The design of any AI systems starts with the choice of training data, which is the first place where unfairness can arise. Training data should sufficiently represent the world in which we live, or at least the part of the world where the AI system will operate. Consider an AI system that enables facial recognition or emotion detection. If it is trained solely on images of adult faces, it may not accurately identify the features or expressions of children due to differences in facial structure.

But ensuring the "representativeness" of data is not enough. Racism and sexism can also creep into societal data. Training an AI system on such data may inadvertently lead to results that perpetuate these harmful biases. One example might be an AI system designed to help employers screen job applicants. When trained on data from public employment records, this system might "learn" that most software developers are male. As a result, it may favor men over women when selecting candidates for software developer positions, even though the company deploying the system is seeking to promote diversity through its hiring practices.[14]

An AI system could also be unfair if people do not understand the limitations of the system, especially if they assume technical systems are more accurate and precise than people, and therefore more authoritative. In many cases, the output of an AI system is actually a prediction. One example might be "there is a 70 percent likelihood that the applicant will default on the loan." The AI system may be highly accurate, meaning that if the bank extends credit every time to people with the 70 percent "risk of default," 70 percent of those people will, in fact, default. Such a system may be unfair in application, however, if loan officers incorrectly interpret "70 percent risk of default" to simply mean "bad credit risk" and decline to extend credit to everyone with that score -even though nearly a third of those applicants are predicted to be a good credit risk. It will be essential to train people to understand the meaning and implications of AI results to supplement their decision-making with sound human judgment.

How can we ensure that AI systems treat everyone fairly? There's almost certainly a lot of learning ahead for all of us in this area, and it will be vital to sustain research and foster robust discussions to share new best practices that emerge. But already some important themes are emerging.

First, we believe that the people designing AI systems should reflect the diversity of the world in which we live. We also believe that people with relevant subject matter expertise (such as those with consumer credit expertise for a credit scoring AI system) should be included in the design process and in deployment decisions.

Second, if the recommendations or predictions of AI systems are used to help inform consequential decisions about people, we believe it will be critical that people are primarily accountable for these decisions. It will also be important to invest in research to better understand the impact of AI systems on human decision-making generally.

Finally-and this is vital -industry and academia should continue the promising work underway to develop analytical techniques to detect and address potential unfairness, like methods that systematically assess the data used to train AI systems for appropriate representativeness and document information about its origins and characteristics.

Ultimately, determining the full range of work needed to address possible bias in AI systems will require ongoing discussions that include a wide range of interested stakeholders. Academic research efforts such as those highlighted at the annual conference for researchers on Fairness, Accountability, and Transparency in Machine Learning have raised awareness of the issue. We encourage increased efforts across the public, private and civil sectors to expand these discussions to help find solutions.

Reliability – AI systems should perform reliably and safely

The complexity of AI technologies has fueled fears that AI systems may cause harm in the face of unforeseen circumstances, or that they can be manipulated to act in harmful ways. As is true for any technology, trust will ultimately depend on whether AI-based systems can be operated reliably, safely and consistently -not only under normal circumstances but also in unexpected conditions or when they are under attack.

This begins by demonstrating that systems are designed to operate within a clear set of parameters under expected performance conditions, and that there is a way to verify that they are behaving as intended under actual operating conditions. Because AI systems are data-driven, how they behave and the variety of conditions they can handle reliably and safely largely reflects the range of situations and circumstance that developers anticipate during design and testing. For example, an AI system designed to detect misplaced objects may have difficulty recognizing items in low lighting conditions, meaning designers should conduct tests in typical and poorly lit environments. Rigorous testing is essential during system development and deployment to ensure that systems can respond safely to unanticipated situations; do not have unexpected performance failures; and do not evolve in ways that are inconsistent with original expectations.

Design and testing should also anticipate and protect against the potential for unintended system interactions or bad actors to influence operations, such as through cyberattacks. Securing AI systems will require developers to identify abnormal behaviors and prevent manipulation, such as the introduction of malicious data that may be intended to negatively impact AI behavior.

In addition, because AI should augment and amplify human capabilities, people should play a critical role in making decisions about how and when an AI system is deployed, and whether it's appropriate to continue to use it over time. Since AI systems often do not see or understand the bigger societal picture, human judgment will be key to identifying potential blind spots and biases in AI systems. Developers should be cognizant of these challenges as they build and deploy systems, and share information with their customers to help them monitor and understand system behavior so that they can quickly identify and correct any unintended behaviors that may surface.

In one example in the field of AI research, a system designed to help make decisions about whether to hospitalize patients with pneumonia "learned" that people with asthma have a lower rate of mortality from pneumonia than the general population. This was a surprising result because people with asthma are generally considered to be at greater risk of dying from pneumonia than others. While the correlation was accurate, the system failed to detect that the primary reason for this lower mortality rate was that asthma patients receive faster and more comprehensive care than other patients because they are at greater risk. If researchers hadn't noticed that the AI system had drawn a misleading inference, the system might have recommended against hospitalizing people with asthma, an outcome that would have run counter to what the data revealed. [15] This highlights the critical role that people, particularly those with subject matter expertise, must play in observing and evaluating AI systems as they are developed and deployed.

Principles of robust and fail-safe design that were pioneered in other engineering disciplines can be valuable in designing and developing reliable and safe AI systems. Research and collaboration involving industry participants, governments, academics and other experts to further improve the safety and reliability of AI systems will be increasingly important as AI systems become more widely used in fields such as transportation, healthcare and financial services.

We believe the following steps will promote the safety and reliability of AI systems:

- Systematic evaluation of the quality and suitability of the data and models used to train and operate AI-based products and services, and systematic sharing of information about potential inadequacies in training data.
- Processes for documenting and auditing operations of AI systems to aid in understanding ongoing performance monitoring.
- When AI systems are used to make consequential decisions about people, a requirement to provide adequate explanations of overall system operation, including information about the training data and algorithms, training failures that have occurred, and the inferences and significant predictions generated, especially.
- Involvement of domain experts in the design process and operation of AI systems used to make consequential decisions about people.
- Evaluation of when and how an AI system should seek human input during critical situations, and how a system controlled by AI should transfer control to a human in a manner that is meaningful and intelligible.
- A robust feedback mechanism so that users can easily report performance issues they encounter.

Creating AI systems that are safe and reliable is a shared responsibility. It is, therefore, critically important for industry participants to share best practices for design and development, such as effective testing, the structure of trials and reporting.

Topics such as human-robot interaction and how AI-driven systems that fail should hand control over to people are important areas not only for ongoing research, but also for enhanced collaboration and communication within the industry.

Privacy & Security – AI systems should be secure and respect privacy

As more and more of our lives are captured in digital form, the question of how to preserve our privacy and secure our personal data is becoming more important and more complicated. While protecting privacy and security is important to all technology development, recent advances require that we pay even closer attention to these issues to create the levels of trust needed to realize the full benefits of AI. Simply put, people will not share data about themselves –data that is essential for AI to help inform decisions about people –unless they are confident that their privacy is protected and their data secured.

Privacy needs to be both a business imperative and a key pillar of trust in all cloud computing initiatives. This is why Microsoft made firm commitments to protect the security and privacy of our customers' data, and why we are upgrading our engineering systems to ensure that we satisfy data protection laws around the world, including the European Union's General Data Protection Regulation (GDPR). Microsoft is investing in the infrastructure and systems to enable GDPR compliance in our largest-ever engineering effort devoted to complying with a regulatory environment.

Chapter 2
Principles, Policies and Laws for the Responsible Use of AI

Like other cloud technologies, AI systems must comply with privacy laws that require transparency about the collection, use and storage of data, and mandate that consumers have appropriate controls so that they can choose how their data is used. AI systems should also be designed so that private information is used in accordance with privacy standards and protected from bad actors who might seek to steal private information or inflict harm. Industry processes should be developed and implemented for the following: tracking relevant information about customer data (such as when it was collected and the terms governing its collection); accessing and using that data; and auditing access and use. Microsoft is continuing to invest in robust compliance technologies and processes to ensure that data collected and used by our AI systems is handled responsibly.

What is needed is an approach that promotes the development of technologies and policies that protect privacy while facilitating access to the data that AI systems require to operate effectively. Microsoft has been a leader in creating and advancing innovative state-of-the-art techniques for protecting privacy, such as differential privacy, homomorphic encryption, and techniques to separate data from identifying information about individuals and for protecting against misuse, hacking or tampering. We believe these techniques will reduce the risk of privacy intrusions by AI systems so they can use personal data without accessing or knowing the identities of individuals. Microsoft will continue to invest in research and work with governments and others in industry to develop effective and efficient privacy protection technologies that can be deployed based on the sensitivity and proposed uses of the data.

Inclusiveness – AI systems should empower everyone and engage people

If we are to ensure that AI technologies benefit and empower everyone, they must incorporate and address a broad range of human needs and experiences. Inclusive design practices will help system developers understand and address potential barriers in a product or environment that could unintentionally exclude people. This means that AI systems should be designed to understand the context, needs and expectations of the people who use them.

The importance that information and communications technology plays in the lives of the 1 billion people around the world with disabilities is broadly recognized. More than 160 countries have ratified the United Nations Convention on the Rights of Persons with Disabilities, which covers access to digital technology in education and employment.

In the United States, the Americans with Disabilities Act and the Communications and Video Accessibility Act require technology solutions to be accessible, and federal and state regulations mandate the procurement of accessible technology, as does European Union law. AI can be a powerful tool for increasing access to information, education, employment, government services, and social and economic opportunities. Real-time speech-to-text transcription, visual recognition services, and predictive text functionality that suggests words as people type are just a few examples of AI-enabled services that are already empowering those with hearing, visual and other impairments.

We also believe that AI experiences can have the greatest positive impact when they offer both emotional intelligence and cognitive intelligence, a balance that can improve predictability and comprehension. AI-based personal agents, for example, can exhibit user awareness by confirming and, as necessary, correcting understanding of the user's intent, and by recognizing and adjusting to the people, places and events that are most important to users. Personal agents should provide information and make recommendations in ways that are contextual and expected. They should provide information that helps people understand what inferences the system is making about them. Over time, such successful interactions will increase usage of AI system and trust in their performance.

Transparency – AI systems should be understandable

Underlying these four preceding values are two foundational principles that are essential for ensuring the effectiveness of the rest: transparency and accountability.

When AI systems are used to help make decisions that impact people's lives, it is particularly important that people understand how those decisions were made. An approach that is most likely to engender trust with users and those affected by these systems is to provide explanations that include contextual information about how an AI system works and interacts with data. Such information will make it easier to identify and raise awareness of potential bias, errors and unintended outcomes.

Chapter 2
Principles, Policies and Laws for the Responsible Use of AI

Simply publishing the algorithms underlying AI systems will rarely provide meaningful transparency. With the latest (and often most promising) AI techniques, such as deep neural networks, there typically isn't any algorithmic output that would help people understand the subtle patterns that systems find. This is why we need a more holistic approach in which AI system designers describe the key elements of the system as completely and clearly as possible.

Microsoft is working with the Partnership on AI and other organizations to develop best practices for enabling meaningful transparency of AI systems. This includes the practices described above and a variety of other methods, such as an approach to determine if it's possible to use an algorithm or model that is easier to understand in place of one that is more complex and difficult to explain. This is an area that will require further research to understand how machine learning models work and to develop new techniques that provide more meaningful transparency.

Accountability

Finally, as with other technologies and products, the people who design and deploy AI systems must be accountable for how their systems operate. To establish accountability norms for AI, we should draw upon experience and practices in other areas, including healthcare and privacy. Those who develop and use AI systems should consider such practices and periodically check whether they are being adhered to and if they are working effectively. Internal review boards can provide oversight and guidance on which practices should be adopted to help address the concerns discussed above, and on particularly important questions regarding development and deployment of AI systems.

Internal Oversight and Guidance – Microsoft's AI and Ethics in Engineering and Research (AETHER) Committee

Ultimately these six principles need to be integrated into ongoing operations if they're to be effective. At Microsoft, we're addressing this in part through the AI and Ethics in Engineering and Research (AETHER) Committee. This committee is a new internal organization that includes senior leaders from across Microsoft's engineering, research, consulting and legal organizations who focus on proactive formulation of internal policies and on how to respond to specific issues as they arise. The AETHER Committee considers and defines best practices, provides guiding principles to be used in the development and deployment of Microsoft's AI products and solutions, and helps resolve questions related to ethical and societal implications stemming from Microsoft's AI research, product and customer engagement efforts.

Developing Policy and Law for Artificial Intelligence

AI can serve as a catalyst for progress in almost every area of human endeavor. But, as with any innovation that pushes us beyond current knowledge and experience, the advent of AI raises important questions about the relationship between people and technology, and the impact of new technology-driven capabilities on individuals and communities.

We are the first generation to live in a world where AI will play an expansive role in our daily lives. It's safe to say that most current standards, laws and regulations were not written specifically to account for AI. But, while existing rules may not have been crafted with AI in mind, this doesn't mean that AI-based products and services are unregulated. Current laws that, for example, protect the privacy and security of personal information, that govern the flow of data and how it is used, that promote fairness in the use of consumer information, or that govern decisions on credit or employment apply broadly to digital products and services or their use in decision-making, whether they explicitly mention AI capabilities or not. AI-based services are not exempt from the requirements that will take effect with GDPR, for example, or from HIPAA regulations that protect the privacy of healthcare data in the United States, or existing regulations on automobile safety.

As the role of AI continues to grow, it will be natural for policymakers not only to monitor its impact, but to address new questions and update laws. One goal should be to ensure that governments work with businesses and other stakeholders to strike the balance that is needed to maximize the potential of AI to improve people's lives and address new challenges as they arise.

As this happens, it seems inevitable that "AI law" will emerge as an important new legal topic. But, over what period of time? And in what ways should such a field develop and evolve?

We believe the most effective regulation can be achieved by providing all stakeholders with sufficient time to identify

Chapter 2
Principles, Policies and Laws for the Responsible Use of AI

and articulate key principles guiding the development of responsible and trustworthy AI, and to implement these principles by adopting and refining best practices. Before devising new regulations or laws, there needs to be some clarity about the fundamental issues and principles that must be addressed.

The evolution of information privacy laws in the United States and Europe offers a useful model. In 1973, the United States Department of Health, Education and Welfare (HEW) issued a comprehensive report analyzing a host of societal concerns arising from the increasing computerization of information and the growing repositories of personal data held by federal agencies. [16] The report espoused a series of important principles -the Fair Information Practices -that sought to delineate fundamental privacy ideals regardless of the specific context or technology involved. Over the ensuing decades, these principles -thanks in large part to their fundamental and universal nature -helped frame a series of federal and state laws governing the collection and use of personal information within education, healthcare, financial services and other areas. Guided by these principles, the United States Federal Trade Commission (FTC) began fashioning a body of privacy case law to prevent unfair or deceptive practices affecting commerce.

Internationally, the Fair Information Practices influenced the development of local and national laws in European jurisdictions, including Germany and France, which in many respects emerged as the leaders in the development of privacy law. Beginning in the late 1970s, the Organization

191

for Economic Coordination and Development (OECD) built upon the Fair Information Practices to promulgate its seminal Privacy Guidelines. As with the HEW's Fair Information Practices, the universal and extensible nature of the OECD's Privacy Guidelines ultimately allowed them to serve as the building blocks for the European Union's comprehensive Data Protection Directive in 1995 and its successor, the General Data Protection Regulation.

Laws in the United States and Europe ultimately diverged, with the United States pursuing a more sectoral approach and the European Union adopting more comprehensive regulation. But, in both cases, they built on universal, foundational concepts and in some cases existing laws and legal tenets. These rules addressed a very broad range of new technologies, uses and business models, as well as an increasingly diverse set of societal needs and expectations.

Today, we believe policy discussions should focus on continued innovation and advancement of fundamental AI technologies, support the development and deployment of AI capabilities across different sectors, encourage outcomes that are aligned with a shared vision of human-centered AI, and foster the development and sharing of best practices to promote trustworthy and responsible AI. The following considerations will help policymakers craft a framework to realize these objectives.

The Importance of Data

It seems likely that many near-term AI policy and regulatory issues will focus on the collection and use of data. The development of more effective AI services requires the use of data -often as much relevant data as possible. And yet access to and use of data also involves policy issues that range from ensuring the protection of individual privacy and the safeguarding of sensitive and proprietary information to answering a range of new competition law questions. A careful and productive balancing of these objectives will require discussion and cooperation between governments, industry participants, academic researchers and civil society.

On the one hand, we believe governments should help accelerate AI advances by promoting common approaches to making data broadly available for machine learning. A large amount of useful data resides in public datasets -data that belongs to the public itself. Governments can also invest in and promote methods and processes for linking and combining related datasets from public and private organizations while preserving confidentiality, privacy and security as circumstances require.

At the same time, it will be important for governments to develop and promote effective approaches to privacy protection that take into account the type of data and the context in which it is used. To help reduce the risk of privacy intrusions, governments should support and promote the development of techniques that enable systems to use personal data without accessing or knowing the identities of individuals.

Chapter 2
Principles, Policies and Laws for the Responsible Use of AI

Additional research to enhance "de-identification" techniques and ongoing discussions about how to balance the risks of re-identification against the social benefits will be important.

As policymakers look to update data protection laws, they should carefully weigh the benefits that can be derived from data against important privacy interests. While some sensitive personal information, such as Social Security numbers, should typically be subject to high levels of protection, rigid approaches should be avoided because the sensitivity of personal information often depends on the context in which it is provided and used. For example, an individual's name in a company directory is not typically considered sensitive and should probably require less privacy protection than if it appeared in an adoption record. In general, updated laws should recognize that processing sensitive information may be increasingly critical to serving clear public interests such as preventing the spread of communicable diseases and other serious threats to health.

Another important policy area involves competition law. As vast amounts of data are generated through the use of smart devices, applications and cloud-based services, there are growing concerns about the concentration of information by a relatively small number of companies. But, in addition to the data that companies generate from their customers, there is publicly available data. Governments can help add to the supply of available data by ensuring that public data is usable by AI developers on a non-exclusive basis. These steps will help enable developers of all types to take greater advantage of AI technologies.

At the same time, governments should monitor whether access to unique datasets (in other words, data for which there is no substitute) is becoming a barrier to competition and needs to be addressed. Other concerns relate to whether too much data is available to too few firms and whether sophisticated algorithms will enable rivals to effectively "fix" prices. All these questions warrant attention; but, they probably can be addressed within the framework of existing competition law. The question of the availability of data will arise most directly when one firm seeks to buy another and competition authorities need to consider whether the combined firms would possess datasets that are so valuable and unique that no other firms can compete effectively. Such situations are unlikely to arise very often given the vast amount of data being generated by digital technologies, the fact that multiple firms often have the same data, and the reality that people often use multiple services that generate data for a variety of firms.

Algorithms can help increase price transparency, which will help businesses and consumers buy products at the lowest cost. But, algorithms could one day become so sophisticated that firms employing them to set prices might establish the same prices, even if the firms did not agree among themselves to do so. Competition authorities will need to carefully study the benefits of price transparency as well as the risk that transparency could over time reduce price competition.

Promoting Responsible and Effective Uses of AI

In addition to addressing issues relating to data, governments have an important role to play in promoting responsible and effective uses of AI itself. This should start with the adoption of responsible AI technologies in the public sector. While enabling more effective delivery of services for citizens, this will also provide governments with firsthand experience in developing best practices to address the ethical principles identified above.

Governments also have an important role to play in funding core research to further advance AI development and support multidisciplinary research that focuses on studying and fostering solutions to the socioeconomic issues that may arise as AI technologies are deployed. This multidisciplinary research will also be valuable for the design of future AI laws and regulations.

Governments should also stimulate adoption of AI technologies across a wide range of industries and for businesses of all sizes, with an emphasis on providing incentives for small and medium-sized organizations. Promoting economic growth and opportunity by giving smaller businesses access to the capabilities that AI methods offer can play an important role in addressing income stagnation and mitigating political and social tensions that can arise as income inequality increases. As governments take these steps, they can adopt safeguards to ensure that AI is not used to discriminate either intentionally or unintentionally in a manner prohibited under applicable laws.

Liability

Governments must also balance support for innovation with the need to ensure consumer safety by holding the makers of AI systems responsible for harm caused by unreasonable practices. Well-tested principles of negligence law are most appropriate for addressing injuries arising from the deployment and use of AI systems. This is because they encourage reasonable conduct and hold parties accountable if they fall short of that standard. This works particularly well in the context of AI for a number of reasons. First, the potential roles AI systems can play and the benefit they can bring are substantial. Second, society is already familiar with a broad range of automated systems and many other existing and prospective AI technologies and services. And third, considerable work is ongoing to help mitigate the risk of harm from these systems.

Relying on a negligence standard that is already applicable to software generally to assign responsibility for harm caused by AI is the best way for policymakers and regulators to balance innovation and consumer safety, and promote certainty for developers and users of the technology. This will help keep firms accountable for their actions, align incentives and compensate people for harm.

Fostering Dialogue and the Sharing of Best Practices

To maximize AI's potential to deliver broad-based benefits, while mitigating risks and minimizing unintended consequences, it will be essential that we continue to convene open discussions among governments, businesses, representatives from non-governmental organizations and civil society, academic researchers, and all other interested individuals and organizations. Working together, we can identify issues that have clear societal or economic consequences and prioritize the development of solutions that protect people without unnecessarily restricting future innovation.

One helpful step we can take to address current and future issues is to develop and share innovative best practices to guide the creation and deployment of people-centered AI. Industry-led organizations such as Partnership on AI that bring together industry, nonprofit organizations and NGOs can serve as forums for the process of devising and promulgating best practices. By encouraging open and honest discussion and assisting in the sharing of best practices, governments can also help create a culture of cooperation, trust and openness among AI developers, users and the public at large. This work can serve as the foundation for future laws and regulations.

In addition it will be critical that we acknowledge the broad concerns that have been raised about the impact of these technologies on jobs and the nature of work, and take steps to ensure that people are prepared for the impact that AI will have on the workplace and the workforce. Already,

AI is transforming the relationship between businesses and employees, and changing how, when and where people work. As the pace of change accelerates, new skills will be essential and new ways of connecting people to training and to jobs will be required.

In Chapter 3, we look at the impact of AI on jobs and work, and offer some suggestions for steps we can take together to provide education and training for people of every age and at every stage of school and their working lives to help them take advantage of the opportunities of the AI era. We also explore the need to rethink protections for workers and social safety net programs in a time when the relationship between workers and employers is undergoing rapid change.

Chapter 2

Principles, Policies and Laws for the Responsible Use of AI

Chapter 3
AI and the Future of Jobs and Work

> **Teachers will not be replaced by technology, but teachers who do not use technology will be replaced by those who do.**

Hari Krishna Arya

Chapter 3
AI and the Future of Jobs and Work

*F*or more than 250 years, technology innovation has been changing the nature of jobs and work. In the 1760s, the First Industrial Revolution began moving jobs away from homes and farms to rapidly growing cities. The Second Industrial Revolution, which began in the 1870s, continued this trend, and led to the assembly line, the modern corporation, and workplaces that started to resemble offices that we would recognize today. The shift from reliance on horses to automobiles eliminated numerous occupations while creating new categories of jobs that no one initially imagined. [17]Sweeping economic changes also created difficult and sometimes dangerous working conditions that led governments to adopt labor protections and practices that are still in place today.

The Third Industrial Revolution of the past few decades created changes that many of us have experienced. For Microsoft, this was evident in how the original vision of our company –to put a computer on every desk and in every home –became reality. That transformation brought information technology into the workplace, changing how people communicate and collaborate at work, while adding new IT positions and largely eliminating jobs for secretaries who turned handwritten prose into typed copy.

Now technology is changing again the nature of jobs and work is changing with it. While available economic data is far from perfect, there are clear indications that how enterprises organize work, how people find work, and the skills that people need to prepare for work are shifting significantly. These changes are likely to accelerate in the decade ahead.

AI and cloud computing are the driving force behind much of this change. This is evident in the burgeoning on-demand -or "gig" -economy where digital platforms not only match the skills of workers with consumer or enterprise needs, they provide for people to work increasingly from anywhere in the world. AI and automation are already influencing which jobs, or aspects of jobs, will continue to exist. Some estimate that as many as 5.1 million jobs will be lost within the next decade; but, new areas of economic opportunity will also be created, as well as entirely new occupations and categories of work. [18]

These fundamental changes in the nature of work will require new ways of thinking about skills and training to ensure that workers are prepared for the future and that there is sufficient talent available for critical jobs. The education ecosystem will need to evolve as well; to help workers become lifelong learners, to enable individuals to cultivate skills that are uniquely human, and to weave ongoing education into full-time and on-demand work. For businesses, they will need to rethink how they find and evaluate talent, broaden the pool of candidates they draw from and use work portfolios to assess competence and skill. Employers will also need to focus more on offering on-the-job training, opportunities to acquire new skills, and access to outside education for their existing workforces.

In addition to rethinking how workers are trained and remain prepared for work, it is important to consider what happens to workers as traditional models of employment that typically include benefits and protections change significantly. The rapid evolution of work could undermine worker protections and benefits including unemployment insurance, workers' compensation and, in the United States, the Social Security system. To prevent this, the legal frameworks governing employment will need to be modernized to recognize new ways of working, provide adequate worker protections, and maintain the social safety net.

The Impact of Technology on Jobs and Work

Throughout history, the emergence of new technologies has been accompanied by dire warnings about human redundancy. For example, a 1928 headline in the New York Times warned that "The March of the Machine Makes Idle Hands." [19] More often, however, the reality is that new technologies have created more jobs than they destroyed. The invention of the steam engine, for example, led to the development of the steam locomotive, which was an important catalyst in the shift from a largely rural and agricultural society to one where more and more people lived in urban centers and worked in manufacturing and transportation -a transformation that

The Future Computed

changed how, when and where people worked. More recently, automated teller machines (ATMs) took over many traditional tasks for bank tellers. As a result, the average number of bank tellers per branch in the United States fell from 20 in 1988 to 13 in 2004.[20] Despite this reduction, the need for fewer tellers made it cheaper to run each branch and allowed banks to open more branches, thereby increasing the total number of employees. Instead of destroying jobs, ATMs eliminated routine tasks, which allowed bank tellers to focus on sales and customer service.[21]

This pattern is common across almost every industry. As one economist found in a recent analysis of the workforce, between 1982 and 2002, employment grew significantly faster in occupations that used computers because automation enabled workers to focus on other parts of their jobs; this increased demand for human workers to handle higher-value tasks that had not been automated.[22]

More recently, public debate has centered on the impact of automation and AI on employment. Although the terms "automation" and "AI" are often used interchangeably, the technologies are different. With automation, systems are programmed to perform specific repetitive tasks. For example, word processing automates tasks previously done by people on typewriters. Barcode scanners and point-of-sale systems automate tasks that had been done by retail employees. AI, on the other hand, is designed to seek patterns, learn from experiences, and make appropriate decisions -it does not require an explicit programmed path to determine how it will respond to the situations it encounters. Together, automation and AI are accelerating changes to the nature of jobs. As one commentator put it, "automated machines collate data -AI systems 'understand' it. We're looking at two very different systems that perfectly complement each other." ²³

As AI complements and accelerates automation, policy-makers in countries around the world recognize that it will be an important driver of economic growth in the decades ahead. For example, China recently announced its intention to become the global leader in AI to strengthen its economy and create competitive advantages.²⁴

Any business or organization that depends upon data and information -which today is almost every business and organization -can benefit from AI. These systems will improve efficiency and productivity while enabling the creation of higher-value services that can drive economic growth. But as far back as the First Industrial Revolution, the introduction of any new technology has caused concern about the impact on jobs and employment -AI and automation are no different. Indeed, it would appear that AI and automation are raising serious questions about the potential loss of jobs in developed countries. A recent survey commissioned by Microsoft found that in all 16 countries surveyed, the impact of AI on employment was identified as a key risk. ²⁵As machines become capable of performing tasks that require complex analysis and discretionary judgment, the concern is it will accelerate the rate of job loss beyond what already occurs due to automation.

While it's not yet clear whether AI will be more disruptive than earlier technological advances, there's no question that it is having an impact on jobs and employment. As was the case in earlier periods of significant technology transformation, it is difficult to predict how many jobs will be affected. A widely quoted University of Oxford study estimated that 47 percent of total employment in the United States is at risk due to computerization. [26] A World Bank study predicted that 57 percent of jobs in OECD countries could be automated. [27] And according to a recent paper on robots and jobs, researchers found that each robot deployed per thousand workers decreased employment by 6.2 workers and caused a decline in wages of 0.7 percent. [28]

Jobs across many industries are susceptible to the dual impact of AI and automation. Here are a few examples: a company based in San Francisco has developed "Tally" which automates the auditing of grocery store shelves to ensure goods are properly stocked and priced[29]; at Amazon, they currently use more than 100,000 robots in its fulfillment centers and is creating convenience stores with no cashiers; in Australia a company has developed a robot that can lay 1,000 bricks per hour (a task that would take human laborers a day or longer to complete); in call centers, they are using chatbots to answer customer support questions; and even in journalism, tasks such as writing summaries of sporting events are being automated. [30]

Even where jobs are not entirely replaced, AI will have an impact. In warehouses, employees have shifted from stacking bins to monitoring robots. In legal environments, paralegals

and law clerks now use "e-discovery" software to find documents. In hospitals, machine learning can help doctors diagnose illnesses more quickly and enable teachers to assess student learning more effectively. But, while AI is changing these jobs, they have not disappeared; there are aspects of the work that simply cannot be automated. Many jobs will continue to require uniquely human skills that AI and machines cannot replicate, such as creativity, collaboration, abstract and systems thinking, complex communication, and the ability to work in diverse environments.

And while it is true that AI will eliminate and change some jobs, it will also create new ones. A recent report from the research firm Forrester projects that by 2027, AI will displace 24.7 million jobs and create 14.9 million new jobs.[31] New jobs will emerge as AI changes how work is done and what people need from the world around them. Many of these jobs will be in technology. For example, banks will need network engineers instead of tellers. Retailers will need people with web programming skills to create online shopping experiences instead of greeters or salespeople on the floor. Farms will need agricultural data analysts instead of fruit pickers. Demand for data scientists, robotics experts and AI engineers will increase significantly.

What's more, AI will create jobs we cannot yet even imagine. While it is relatively easy to see where automation may reduce the need for workers, it is impossible to foresee all of the changes that will come. As one economic historian put it, "we can't predict what jobs will be created in the future, but it's always been like that."[32]

One result of the rapid transformation of work caused by AI and automation is a shortage of critical talent across many industries. As jobs increasingly require technology skills, companies compete for the employees who have specialized skills supporting digital capabilities such as robotics, augmented reality computations, cybersecurity and data science. It is estimated that by 2020, 30 percent of technology jobs will go unfilled due to talent shortages[33], and this gap is likely to widen given the time it takes to introduce training programs for new technology skills. According to the World Economic Forum, many academic fields experience unprecedented rates of change in core curriculum. They suggest that nearly 50 percent of subject knowledge acquired during the first year of a four-year technical degree will become outdated before students graduate. And by 2020, more than a third of the skills needed for most occupations will be ones that are not considered crucial today.[34] More broadly, technology will significantly impact the skills requirements in all job families. To manage these trends successfully, we'll need to ensure that people in the workforce can continually learn and gain new skills.

Economists who are studying the emerging talent shortage and the replacement of so-called "middle skills" jobs by automation worry that technological advances such as AI are widening the income gap between those with technological skills and training and those without.[35] As expertise in areas such as data analytics becomes more central to many jobs and automation enables machines to handle more repetitive tasks, demand for highly skilled workers will grow, and the need for those with fewer skills will fall -an effect known as the "skill-

The Future Computed

biased technical change." For example, while the number of jobs for Americans with a four-year college degree doubled between 1989 and 2016, the job opportunities for those with a high school diploma or less fell by 13 percent. Over the same period, the number of Americans with a college degree grew by just under 50 percent and the unemployment rate for those without a college degree rose 300 percent compared to those with a college degree.[36] Addressing this widening gap will require a shift in how we think about education and training so that we can prepare more of the workforce to take advantage of the opportunities that are emerging.

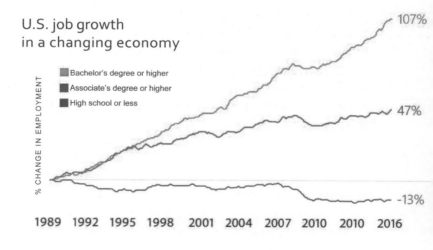

Chart 6.
Source: Georgetown Center on Education and the Workforce

The Changing Nature of Work, the Workplace and Jobs

Chapter 3
AI and the Future of Jobs and Work

Until recently, most people worked in traditional employer-employee relationships at specific worksites: offices, factories, schools, hospitals or other business facilities. This traditional model is being upended as more people are engaged through remote and part-time work, such as contractors, or through project-based employment.

Some studies have noted that between 2005 and 2015, the number of people in alternative work relationships -which include contractors and on-demand workers -increased from 10 percent to 16 percent accounting for nearly all net job growth during that period.[37] A recent study by the McKinsey Global Institute concluded that "the independent workforce is larger than previously recognized" with up to 162 million people in Europe and the United States -20 or 30 percent of the working-age population -engaged in some form of independent work. For more than half of these individuals, independent work supplements their primary source of income.

These alternative work arrangements are fueled by advances in technology. Perhaps the most notable trend in this regard is the rise of the on-demand economy. At its core, the on-demand economy refers to working arrangements in which people find work through online talent platforms or staffing agencies, performing tasks for a wide variety of customers. According to the McKinsey Global Institute, 15 percent of independent workers use digital talent platforms to connect to

The Future Computed

work. Researchers at Oxford University's Martin Programme on Technology and Employment estimate that nearly 30 percent of jobs in the United States could be organized into task-based work within 20 years. [38]

The on-demand economy presents enormous opportunities for workers and businesses. McKinsey estimates digital platforms that match workers with opportunities could raise global GDP by as much as 2 percent by 2025, increasing employment worldwide by 72 million full-time equivalent jobs. Here is just a partial list of the potential benefits of the on-demand economy:

▸ Engagement in on-demand work through digital platforms allows jobs to come to workers, rather than forcing people to migrate to available work. This helps workers who live in areas where job opportunities are limited and enables companies to access a wider talent pool.

▸ According to the Hamilton Project, more than 70 percent of labor force non-participants report that caregiving, disability or early retirement keeps them out of the workforce. The flexibility of on-demand work reduces the barriers that traditional employment models present. [39] According to a survey by the Pew Research Center, nearly 50 percent of on-demand workers report a "need to control their own schedule." Another quarter said there was a "lack of other jobs where they live." [40]

▸ The on-demand economy offers more opportunities for part-time labor. Today, many workers prefer the flexibility of part-time work to full-time employment.⁴¹ For millennials, flexibility, work/life balance, and the social impact of their work can be more important than a high salary or a full-time career. And many baby boomers are choosing to work later in life, often through part-time work.

- ▸ The on-demand economy allows businesses to engage workers on a short-term basis, facilitating business agility and reducing long-term staffing costs. The on-demand economy can be particularly helpful to small businesses that cannot afford a large full-time workforce but can get work done through targeted on-demand engagements. Costs can be reduced further by recruiting freelancers through online platforms that feature competitive bids for projects.
- ▸ The on-demand economy can provide companies with access to skills they do not have in-house. Hiring freelancers enables employers to find individuals with specific skills and engage them on an as-needed basis.
- ▸ The on-demand economy provides access to supplemental income. For instance, the online platform Teachers Pay Teachers includes an online marketplace where teachers buy and sell lesson plans and other educational resources.⁴²

While the on-demand economy has the potential to promote greater labor force participation, many concerns have been raised about its impact on working conditions and worker protections. Some of these concerns include:

- Because the on-demand economy is so new, it is stretching the bounds of existing regulations relating to worker protections, including child labor laws and minimum wage requirements. While some on-demand digital platforms offer worker protections, others have taken the position that even baseline worker protections do not apply to the on-demand labor model.

- The on-demand borderless workplace heightens issues relating to wages and the distribution of the global workforce. Because of the differences in the cost of living across the globe and the opportunity for employers to hire workers where wages are low, jobs may move from the higher-wage to the lower-wage countries.

- Some studies have shown that the economic benefits of the on-demand economy largely accrue to platform owners and consumers, but not to workers.[43] Because these platforms commoditize work into tasks, they may devalue other contributions that workers can make to the platform or the overall digital economy.

- The commoditization of the workforce also has the potential to reduce access to social insurance, career development and social interaction, which might otherwise strengthen innovation and economic value. Moreover, workers in the on-demand economy do not benefit from the investments enterprises make in work culture.

- In the long term, as platforms "learn" from workers and automate more tasks, the development of the platform economy may contribute to the elimination of jobs. Those who are unable to acquire new skills may be marginalized, further concentrating wealth in the hands of platform owners and top earners.

As the on-demand economy continues to grow, enterprises have an opportunity to shape policy within their own companies, at the industry level and from a public policy perspective. Increasingly, the technology industry needs to engage to change the perception that it reaps the benefits of technology progress at the expense of workers who are displaced or left without protections, benefits or long-term career paths.

Companies must acknowledge the impact of the on-demand model on workers rather than claim that they are "just the technology platform." Companies that do not acknowledge the importance of worker protections and benefits risk damage to their brands and face the possibility that lawmakers and the courts will step in to impose regulations that could limit the business opportunities that the on-demand economy presents. Microsoft believes that companies can benefit from the on-demand economy while taking steps to provide protections, benefits and opportunities that offer long-term economic stability for workers.

The technologies underpinning the on-demand economy are also changing how enterprises organize work within their traditional workforce. Today, a wide range of factors are driving enterprises to focus on creating a globally distributed workforce, including the need to look beyond local talent pools to find people with the skills that they need. But, as countries face nationalist pressures and businesses face more restrictive immigration laws, companies may also need to consider expanding their domestic workforce.

New technologies and tools are enabling businesses to accommodate distributed workforces. Online platforms can aggregate data on workers and job openings across entire countries and regions, making it easier to address geographic mismatches between skills and jobs. And because new collaboration tools support remote work, employees are no longer tied to working in a fixed location. In addition, people are seeking more flexibility in how and where they work. In a recent poll, 37 percent of technology professionals said they would take a 10 percent pay cut to work from home. [44]

While the new technologies are allowing businesses to distribute work across the globe, they require shifts in the way enterprises train workers, cultivate culture, and build institutional knowledge and intellectual property. Today, many enterprises are finding that more dispersed workforces make effective collaboration more difficult and agility more challenging. As the unit of work shifts to task-based projects that use new agile team structures, the combination of alternative employment arrangements and distributed workers means enterprises need to reconsider how they engage employees, build teams, and support career development and training. Enterprises will need to take advantage of collaboration tools like Microsoft Teams or Slack to address these shifts. They will need to use learning platforms like LinkedIn Learning or Coursera to address employees' needs for career development and mentorship. In addition, they will need to discover news ways to build community and engagement within a dispersed workforce.

Preparing Everyone for the Future of Work

Because the skills required for jobs in the AI economy are changing so rapidly, we need to ensure that our systems for preparing, educating, training, and retraining the current and future workforce also evolve. Not only will the new AI economy require new technical skills, but there is a growing recognition that most workers will need to learn new skills throughout their working lives.[45]

According to a recent study by the Pew Research Center, 87 percent of U.S. adults in the labor force say that to keep up with changes in the workplace, it will be essential or important to get training and develop new skills throughout their working lives. The ability to learn new things, collaborate, communicate and adapt to changing environments may become the most important skills for long-term employability. Innovation and new solutions throughout our education, training and workforce systems will be required to help people stay competitive in this rapidly changing workforce.

As automation and AI take on tasks that require thinking and judgement, it will become increasingly important to train people -perhaps through a renewed focus on the humanities -to develop their critical thinking, creativity, empathy, and reasoning.

Employers have a responsibility to help the education and workforce systems better understand, interpret and anticipate what professional skills they'll need. While we can't predict with certainty which jobs will exist in the future, we believe strongly that education and training will be more important than ever. Technology can be better utilized throughout the system to help students and job seekers discover promising career paths, assess their current skills, develop new skills and connect to jobs, and to scale the solutions to meet the needs of broader swaths of the population.

For people to succeed in the age of automation and AI, improving education and training systems for everyone will be critical. Most experts agree that some post-secondary education and training will be essential. The following charts show the clear relationship between educational attainment and employment levels. Chart 7 reflects this strong positive relationship in OECD countries. Chart 8 shows the United States unemployment rate impacts those with less education disproportionately and more acutely than those who accrue more education.

The stark differences in the increases in unemployment rate, particularly for those with less education, demonstrate a higher volatility to that group. This is yet another example of how technology companies can play a vital role in shaping education and labor policy.

To help people get the training they need to thrive in today's economy and prepare for the future, Microsoft is focusing on three areas: 1) preparing today's students for tomorrow's jobs; 2) helping today's workers prepare for the changing economy; and 3) creating systems to better match workers to job opportunities.

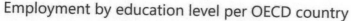
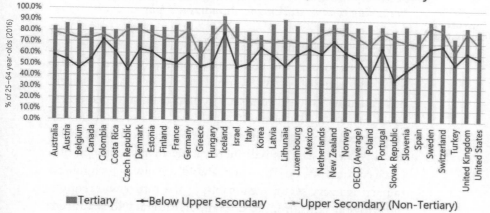

This chart shows that in the U.S., this relationship between education and employability strengthens over time as the workforce requires even more skilled workers.

Chart 7.
Source: OECD, Employment by education levels, Percentage of 25-64-year-olds, 2016.

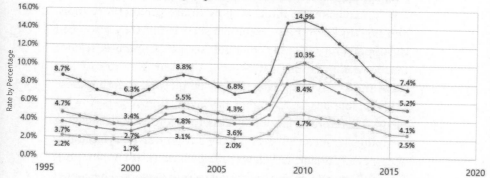

Chart 8.
Source: U.S. Bureau of Labor Statistics

The Future Computed

Preparing today's students for tomorrow's jobs

The single most important skill that people will need for tomorrow's jobs is the ability to continually learn. Future jobs will require what Stanford professor Carol Dweck has called a "growth mindset" to engage in more complex problem-solving. Success will require strong communication, teamwork and presentation skills. People will need to be more globally aware as jobs will increasingly involve serving not just a community, but the world. Rapidly evolving technology impacting every sector means jobs of the future will require more digital skills, from basic computer literacy to advanced computer science. And emerging technologies and the jobs of the future will require more digital and computer skills.

Given these changing expectations, the skills young people need to learn before entering the workforce have also changed. Every young person needs to understand how computers work, how to navigate the internet, how to use productivity tools, and how to keep their computers secure. But they also need the opportunity to study computer science. Computer science teaches computational thinking, a different way to problem solve and a skill in high demand by employers. Together these skills enable access to higher paying jobs in faster-growing fields. Therefore, equitable access to rigorous and engaging computer science courses must be a top priority. If equitable access is left unaddressed, we will exclude entire populations from fully participating in this new world of work. The goal of equitable access should be computer science classrooms that are diverse across race, gender, disability and socioeconomic status.

Some countries, such as the United Kingdom, embed instruction in computational thinking into classes at every grade level, while others struggle to close the digital skills and computer science education gap. For example, while the United States has made progress to ensure that all students can take at least one computer science class before graduating from high school, thousands of students still do not have access.[46] According to the College Board, last year only 4,810 of the 37,000 high schools in the United States offered the Advanced Placement computer science exam. with girls, minorities, and the economically disadvantaged least likely to have access.[47]

To help address the global need for digital skills development, Microsoft Philanthropies partners with governments, educators, nonprofits, and businesses and is involved in a range of programs and partnerships aimed at addressing the skills gap at scale. Together with our partners, we're working to help prepare young people for the future, especially those who might not otherwise have access to opportunities to acquire critical skills. For example, through our YouthSpark program, Microsoft works with 150 nonprofit organizations in 60 countries to offer computer science learning both in and out of school to more than 3 million young people.

Microsoft Philanthropies partners with governments, educators, nonprofits, and businesses to help address this gap. We work with 150 nonprofit organizations in 60 countries to offer computer science learning, both in and out of school. To date we've reached more than 3 million youth, 83 percent of whom are from underserved communities and more than half are female.

To solve this problem, increasing the number of teachers who are trained to teach computer science is also critical. Technology Education and Literacy in Schools (TEALS) is a program that operates in 349 high schools in 29 states throughout the United States and is supported by Microsoft Philanthropies. The program engages 1,000 tech volunteers from over 500 different companies to team-teach computer science, usually with the math or science teacher. Within two years of working with their volunteer, 97 percent of classroom teachers are able to teach computer science on their own, creating the basis for sustainable computer science programs.

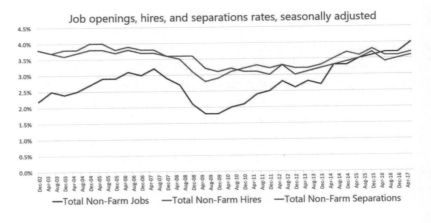

Chart 9.
Source: U.S. Bureau of Labor Statistics, Job Openings and Labor Turnover Survey, October 2017.

Supporting today's workers for the changing economy

Because technology is changing so rapidly, it's not enough to just focus on educating tomorrow's workforce; we must also help today's workers gain skills that are relevant in the changing workplace. Economic growth depends on a skilled workforce that can enable enterprises to take advantage of a new generation of emerging digital technology innovations. To achieve this, workers will need to be lifelong learners. As noted earlier, the global economy is going through rapid changes as automation and AI create demand for a more skilled workforce. This is reflected in recent labor statistics in the United States where, for the first time, job postings have surpassed hiring in the monthly U.S. Bureau of Labor Statistics Job Openings and Labor Turnover Survey (JOLTS) reporting. [48]

This is just one illustration of the global mismatch between employer needs and the skills that today's workers possess. According to a 2017 survey by global staffing firm ManpowerGroup, significant skills shortages exist in Japan, India, Brazil, Turkey, Mexico, Greece, Australia and Germany. [49] In the United States, the National Skills Coalition reports that 53 percent of jobs today are "middle skill" or "new-collar" jobs that require more than a high school diploma and less than a college degree. But only 43 percent of the workforce is a match for this requirement. At the same time, while 20 percent of the workforce has a high school graduation credential or less and is considered "low-skilled," just 15 percent of jobs are open to people with this level of educational attainment.[50] Further, in a study of job postings

Chapter 3
AI and the Future of Jobs and Work

by Burning Glass Technologies, 8 out of 10 middle-skills jobs require basic digital literacy skills, something that more than half of workers today lack. Unless we change how we prepare people for these new jobs, this gap will continue to widen.[51] The National Skills Coalition predicts that 80 percent of jobs that will be created by 2024 will require post-secondary credentials.[52]

As demands for a more educated and skilled workforce continue to grow, we must identify new ways to increase the skills of today's workers. Workforce systems will need to evolve to keep pace with the changing technologies. Emerging practices focused on distance and online learning as well as investment in more on-the-job training programs are key ways to prepare today's workers for the changing workplace.

To understand how to train the current workforce, it is important to identify the skills that enterprises need. Microsoft and its LinkedIn subsidiary are already experimenting with new ways to understand which skills are currently in demand and how to help people gain them.[53] For example, LinkedIn is working with the National Cybersecurity Center (NCC) and the University of Colorado at Colorado Springs to identify the most in-demand cybersecurity occupations in the United States and map the skills needed to be hired for those jobs.

LinkedIn is also working with local training programs to update curriculum and to teach graduating students how to use LinkedIn in their job search. Microsoft offers curriculum and certification programs to help people develop digital skills through programs like Imagine Academy, YouthSpark and LinkedIn Learning.[54] This is important because digital skills are critical in all job clusters. In fact, research firm IDC reports that knowing how to use Microsoft Office was the third most cited skill requirement across all occupations.[55]

It will also be critical to identify new entry points into the workforce. As enterprises face talent shortages, they must explore new ways to bring in talent from available labor pools. Microsoft and LinkedIn are testing several programs that do this, such as Microsoft Software and Systems Academy (MSSA). An 18-week training program, MSSA is created specifically to prepare service members and veterans for careers in cloud development, cloud administration, cybersecurity administration, and database and business intelligence administration. At the end of the program, graduates interview for full-time jobs at Microsoft or one of our hiring partners. So far, 240 companies have hired graduates of MSSA. Microsoft is also working with the state of Washington's Apprenticeship and Training Council, which offers the first registered apprenticeship program for the IT industry.

LinkedIn supports the apprenticeship model as well, and is working to identify ways to build an apprenticeship marketplace. LinkedIn launched REACH, a six-month apprenticeship program where people join a LinkedIn engineering team to learn what it is like to work as a software engineer and gain experience to help them pursue a career in software development. LinkedIn is also partnering with CareerWise Colorado to create a marketplace that lists apprenticeship openings for high school students. And it is working with the state of Colorado's apprenticeship office to help people understand the value of apprenticeships.

All of these programs are good first steps. But the next -and maybe harder -challenge will be to figure out how to scale these programs through public and private partnerships to have sustainable impact on the workforce. This will require both educational institutions to think differently -at scale -about how they train, and employers to think differently about how they identify and onboard talent.

Supporting the development of systems to create a skills-based marketplace

To help foster economic prosperity across the globe, the public and private sectors must also invest in new educational delivery models. People need to be able to gain the demand-driven skills required for advancement and a system needs to be developed in which credentials are portable, stackable and valued by employers. The rapidity of change in the workplace requires employers and workforce providers to work together in new ways. The public and private sectors should seek to meet the needs of people at all stages of the workforce continuum -from students entering the workforce to unemployed and underemployed workers, to people currently in the workforce who need help gaining new skills to ensure their long-term employability.

To help enterprises find qualified employees and workers find jobs, we'll need to shift from a system based on traditional degrees to a system based on skills. This system should account for the rapidly evolving skills employers need across occupations. And it should recognize the skills individuals possess to more efficiently connect workers to employers.

A first step will be to create a common taxonomy of skills. Emerging technologies and changes to the workplace require education providers to offer training in the skills that employers need. It will be critical to codify the most in-demand skills and train workers on them and on how to articulate their skills to potential employers.

Employers and workforce agencies should use real-time labor market information to identify in-demand skills, a task for which LinkedIn and the broader IT industry are well-placed to assist governments and workforce agencies. Governments can use this information to develop and deliver high-quality workforce training programs and offer incentives and financial resources to private and nonprofit organizations to provide training. Goals for educational attainment should include outcomes related to employment, skills and advancement.

Technology and data must be used to build a dynamic skills-based labor marketplace that guides the education and workforce systems. To be successful, we'll need a worker-centered framework for assessing learning outcomes that harmonizes data across sectors in a way that is easier for individuals to navigate. This framework should emphasize the knowledge that employers require and include the technical and foundational skills workers need in the digital workplace. Foundational skills include problem-solving, work ethic, teamwork, curiosity and interpersonal communication. This framework should guide training organizations as they help people acquire skills and earn credentials.

We also need to identify existing open jobs and the skills required to fill them. Digital platforms such as LinkedIn, TaskRabbit and Upwork offer insights about in-demand skills based on job or task openings. Over time, this data can be used to construct analyses such as the LinkedIn Economic Graph to understand supply and demand for specific skills and how they vary over time for a given region, particularly when combined with government data on local demographics and businesses.

Microsoft and LinkedIn are taking additional steps to understand which skills are in high demand, to invest in skills development to address the changing nature of work and of jobs, and to help people find jobs to match their skills. In pursuit of these goals, Microsoft and LinkedIn have partnered with Skillful, an initiative of the Markle Foundation, that is creating a skills-based labor market that works for everyone, with a focus on those without a college degree. Microsoft has made a substantial investment to help the Markle Foundation build this marketplace. [56]

To achieve its mission, Skillful helps employers expand their talent pool by providing data, tools and resources to simplify the adoption of skills-based practices. Coaches and digital services enable job seekers to find out which skills are in demand and access professional training at any stage of their careers. Skillful also works with educators and employers to ensure that students are learning the skills they need to succeed in today's digital economy. The partnership aims to create a model that can be replicated across the United States, aiming to help millions of Americans find rewarding careers.

Skillful is also working with LinkedIn to test strategies to improve the lives of skilled workers through initiatives such as Mentor Connect, LinkedIn's pilot mentorship program that uses Skillful's coaching efforts and platform.

To improve how the public and private sectors work together to match job seekers with job openings, LinkedIn has opened its listings to governments in the United States, free of charge. In 2017, more than 1 million government job listings appeared on LinkedIn. In addition, the National Labor Exchange, which is managed by the National Association of State Workforce Agencies and includes jobs from all 50 state job banks, began sending its jobs to LinkedIn in January 2017. LinkedIn has shared labor market insights with more than 70 U.S. cities through the White House TechHire program. LinkedIn has also shared data with government agencies in New York, Los Angeles, Chicago, Louisville, New Orleans, Seattle, San Francisco and Cleveland to help them improve issues such as student retention and youth unemployment, identify job biases, and understand supply and demand for job skills.

While many of these programs are relatively new, it is clear that we need to use data to build a more dynamic skills-based labor marketplace that guides education and workforce systems and prepares workers for available jobs.

Changing Norms for Changing Worker Needs

To meet the challenges of the evolving economy, we must also understand how the on-demand economy, part-time work, independent contracting, and temporary jobs affect individuals and society.[57] These changes raise questions that are not always adequately addressed by existing legal and policy frameworks.

To enable innovation and to protect workers, the public and private sectors must tackle a number of key policy questions. Legal certainty must be created so that workers and businesses understand their rights and obligations. Industry must also define its own standards for worker protection to ensure that society does not become further divided between the "haves and have-nots." To promote the efficient flow of skills, encourage entrepreneurship, and allow workers to exercise their market power to the best of their ability, industry and governments must work together to find ways to enable workers to take their benefits with them as they change jobs. And the social safety net must be modernized to support workers and families, as well as stabilizing the economy during periods of economic instability and labor market shifts.

Providing Legal Certainty and Structure for Employers and Workers

Given the velocity of change in the modern workforce, it is not surprising that existing legal and policy frameworks do not adequately address all of today's changing work arrangements.

Questions and uncertainty about how to categorize workers have been an issue for a sometime, with consequences for businesses, workers and government. Now, changes in labor marketplaces and the rise of on-demand work platforms are increasing the urgency to find answers to these questions.[58]

Broadly speaking, current laws tend to recognize only two designations for workers: 1) employees who work on a regular basis in a formal relationship with an employer; or 2) independent contractors who provide goods or services under a specified contract.

Employees have traditionally enjoyed less flexibility and control over their hours and working conditions, but retain more stability and legal protection. Independent contractors typically retain more control over when and how they work, but receive fewer legal protections. Whether or not someone is an employee determines whether they are protected by traditional labor, wage and hour, and equal opportunity laws, and whether they can access employer-provided benefits such as private pensions, access to training, retirement benefits and, in many countries, healthcare. A worker's designation also determines whether employers contribute to and workers benefit from social safety net benefits such as unemployment insurance and, in the United States, Social Security and state-paid leave benefits.

Today, most on-demand workers are treated as independent contractors by digital platforms and the businesses that engage them. Under this classification, on-demand workers are not protected by minimum wage and overtime pay requirements, child labor regulations, or anti-discrimination and anti-harassment laws. In addition, there is a lack of clarity about the rights and protections that workers who connect through an intermediary can expect under the law. As dissatisfaction about the lack of protections grows, on-demand workers are increasingly challenging such designations through litigation or government intervention. [59]

The results have been inconsistent. For on-demand workers, this creates uncertainty about what rights and benefits they can expect. For platform companies and the businesses that engage on-demand workers, it raises questions about whether on-demand workers will be considered to be employees, subject to the associated costs and protections.

Until labor and employment laws and systems for providing benefits are modernized to respond to current workforce trends, there's a danger that growth in productivity and opportunity will be constrained. There is a risk that if we fail to impose baseline protections - including wage protection -work will become increasingly stratified between high paying, stable employment and low-value, low-paid, task-oriented gigs. This may undermine the potential of the on-demand economy. Unfortunately, current discussions about the classification of workers are often extremely polarized -with business pushing for narrower classifications and labor advocates pushing for more expansive interpretations. What is needed is broader dialogue about the needs of businesses and workers to determine what changes are required to serve the interests of both in a way that is productive and fair.

So far, policy recommendations have focused on either redefining the categories of employees and independent contractor or finding ways to mitigate the consequences of this difference -often by extending protections, benefits and social safety net participation to contingent workers. Both of these approaches focus on addressing the issue by making the distinction between the two categories less extreme and providing basic protections to workers who are currently left out. Current policy proposals include the implementation of a new worker classification for "independent worker" that would fall between employee and independent contractor; the creation of a safe harbor for income and employment tax purposes for certain workers; the expansion of collective bargaining and other protections to certain classifications of on-demand workers; and the adoption of voluntary minimum industry standards for worker protections. All of these proposals should be explored as more people find work through on-demand platforms.

Developing Industry Standards to Protect All Workers

Today, business leaders have the opportunity to play a significant role in reshaping employment policy for the emerging economy by setting their own standards for on-demand engagements. Microsoft believes we can (and should) positively impact the treatment of on-demand workers through its internal policy. Microsoft's policy includes minimum pay requirements for all on-demand work. It requires that on-demand workers be paid within one week of completing work and that all workers be treated with dignity and respect. It also prohibits the use of child labor and requires the on-demand platforms that it uses to be accessible. Microsoft is implementing contractual terms with the on-demand platforms it engages with that reflect this policy.

While corporate policies can provide some degree of protection to on-demand workers, the impact is limited. But enterprise users of on-demand labor also have an opportunity to contribute to broader solutions to these issues. For instance, groups such as freelancers' unions and caregiver coalitions have improved standards for task workers -sometimes through legislation. Approaches like the National Domestic Workers Alliance's "Good Work Code" for domestic workers in the United States offer a framework for engaging workers that includes safety, shared prosperity, a living wage, inclusion and input.[60] Industry leaders should encourage discussions among businesses and workers to develop standards like these for task-based work that might include wage, benefits and fair treatment commitments. This could lead to a set of standards endorsed by businesses that might serve as a framework for nongovernmental policy. Such standards could either be industry-specific or generalized to broader platforms, and might also serve as the framework for legislation that sets minimum protections.

Ensuring Benefits Move with Workers

These labor market trends have tremendous implications for both worker protections and employer-provided benefits. The employer-based benefits model that emerged in most of Europe and North America in the middle of the last century is based on two principles: first, that businesses benefit from the well-being of a stable workforce; and second, that certain benefits are best provided by employers rather than the government as an investment in workforce stability.

This approach has shaped our perspective of the social contract between employers and employees. While the nature of work has evolved with technology innovation, the system of employer-provided benefits and social safety nets has not. The challenge we face now is how to transform benefits and social insurance programs to provide adequate coverage for workers and a sustainable contribution structure for businesses.

In today's digital economy, the mobility of labor and the ability to quickly focus skills on new growth areas are vitally important to business success. Many businesses may find the relative burden of maintaining employer-provided benefits not worth the cost. Individual workers also want benefits that are portable and flexible. Portability of benefits will be critical to a viable solution. Three models have emerged as possible solutions.

▸ **Employer-provided benefits.** The issue of providing benefits to people working in industries that are structured around short-term projects is not new. Industries such as construction and entertainment have addressed this through labor-management partnerships that enable workers to retain healthcare and pension coverage across multiple employers, even for short-term work. A collective bargaining structure provided a way for employers to contribute to benefits pools without bearing the burden of administration; workers did not have to be responsible for moving benefits and seeking out new providers. New models could use this approach, which would reduce inefficiency and confusion, and ensure that workers have access to basic protections and adequate benefits. This would support greater labor mobility because workers would be less likely to stay in jobs simply to retain benefits.

- **Use of new platforms to provide benefits.** The rise of on-demand labor platforms may create opportunities to develop new ways for workers to access benefits. For example, Care.com, a platform for caregivers, enables families to contribute to their caregiver's benefits in a way that is similar to how traditional corporate employers fund employee benefits.[61] When families pay a caregiver through Care.com, a percentage funds benefits that stay with caregivers even when they go to work for other families on Care.com. There are still challenges to this approach -including what happens when workers find work through different platforms.

- **Government mandates and funds.** In some countries, national or even multinational government organizations may seek to address this gap. In those countries where a broad new nationwide system may not be feasible, smaller governmental units may be able to establish the infrastructure and risk pooling needed to make benefits affordable. Some countries require basic benefits, with an accompanying structure to provide those benefits. In the United States, where broad new federal programs have not received political support, some states have sought to create their own healthcare or retirement programs. In the short term, policymakers should consider creating pilot programs to establish portable benefits, such as, legislation introduced at the state level in the United States. [62]

Modernizing the Social Safety Net

A more mobile and dynamic workforce will increase pressure on social safety net programs. As people find work through a more diverse array of non-exclusive arrangements that may not include employer-provided benefits or allow workers to earn enough to build their own savings, they will rely more than ever on safety net programs like unemployment insurance, workers' compensation and Social Security.

Programs that are triggered during a worker's productive working years are particularly important for workers' economic stability, which in turn helps maintain a diverse and skilled workforce. Periods of joblessness produce income volatility, which can have serious long-term consequences for workers and their families. This also reduces the pool of available skilled labor for businesses. Even a robust economy includes a significant level of under-employment or unemployment. In August 2017, the U.S. Bureau of Labor Statistics estimated that 7.1 million American workers were unemployed, with an additional 5.3 million working part-time for economic reasons or as involuntary part-time workers. These periods are likely to occur many times over a worker's life.[63]

Many existing social safety net programs are already underfunded and face further fiscal pressures as workforces age. This means that during periods when there is increased need, such as during a recession, existing safety nets are likely to prove inadequate. Compounding the problem, many comprehensive safety net programs are heavily dependent on traditional employment relationships. A significant shift away from traditional employment without corresponding policy changes could further erode work-based social safety net programs. Finally, these programs do not take into account newer models of work, nor do they anticipate that individuals may move in and out of the workforce with greater frequency or for a greater variety of reasons. It will be essential to modernize these programs to encourage labor mobility and enable workers to gain new skills and connect to new opportunities.

Companies can begin to experiment with public-private partnerships to explore how to meet the needs of workers. For example, Microsoft, through LinkedIn, is exploring new ways to speed the re-employment of workers in the United States. LinkedIn is working with the state of Utah to test network-based job searching as a strategy for reemployment through a pilot program that was recently highlighted by the Trump administration for saving taxpayer money by enabling unemployed workers to find new jobs more effectively. In addition, Microsoft and LinkedIn are building tools for employment counselors and job seekers that would improve workforce programs such as unemployment insurance and state workforce programs. And LinkedIn is working with the National Association of State Workforce Agencies to produce job search curriculum for its network of 2,500 publicly managed job centers in the United States.

Enterprises should continue to use data and technology tools to assist governments in identifying opportunities for worker redeployment to scale these solutions beyond pilots and experiments. However, modernizing the social safety net will require a multifaceted approach such as:

- **Rethink unemployment insurance and reemployment programs, including job training and trade adjustment assistance programs.** Steps have been proposed to begin modernizing unemployment insurance and to bolster the program's solvency. Businesses should engage in discussions about the importance of next-generation versions of unemployment insurance and employment services that take into account newer models of work; anticipate that individuals may move in and out of the workforce with greater frequency; promote greater labor mobility; and help workers gain new skills and connect with new opportunities.

- **Reform tax policy and social safety net.** Policymakers must explore how to adjust policies to adequately fund social safety net programs. This may include going beyond existing tax bases to consider other methods of funding social safety nets. For example, some have questioned whether wages are the right measure of income to be taxed. Where business productivity may be better measured by production than through wages, some propose assessing taxes to support social safety nets and government revenue based on other measures.

The case must also be made for how social programs can increase the size of the labor pool; be structured to help employees move in and out of work more easily and more flexibly; and reduce burdens for employers. Without significant modernization, social safety nets will not adequately support emerging models of work. The private and public sectors must join together to explore how to best support workers in the new economy.

Working Together

As we move forward, it will be essential for governments, the private sector, academia, and the social sector to join together to explore how to best support workers in the new economy. This can be achieved by developing new approaches to training and education that enable people to acquire the skills that employers need as technology advances; by creating innovative ways to connect workers with job opportunities; and by modernizing protections for workers to promote labor mobility and cushion workers and their families against uncertainty in a fast-changing global economy.

Conclusion
AI Amplifying Human Ingenuity

Conclusion
AI Amplifying
Human Ingenuity

What happens when we begin to augment human intelligence and ingenuity with the computational intelligence of computers? What does human-centered AI look like?

It may look a lot like Melisha Ghimere, a 20-year-old computer science student at Kantipur Engineering College in Kathmandu, Nepal. Melisha's team was a regional finalist in Microsoft's Imagine Cup competition in 2016.

Like the vast majority of the people of Nepal, she comes from a family of subsistence farmers who raise cows, goats and water buffalo. Over the years, her aunt and uncle, Sharadha and Rajesh, did well, building a herd of more than 40 animals -enough to raise two children, support four other relatives, and even hire a few workers to help out. But then, seven years ago, an outbreak of anthrax wiped out much of their herd. They are still struggling to regain their economic footing.

At college, Melisha's family was never far from her mind. So she set out to develop a technology-based solution that would help farmers like her uncle. Working with three other students, she researched livestock farming and veterinary practices, and spoke with many farmers. Together, they built a prototype for a monitoring device that tracks temperature, sleep patterns, stress levels, motion and the activity of farm animals. Melisha's AI system predicts the likely health of each animal based on often subtle changes in these observations. Farmers can follow the health of their animals on their mobile phones, access advice and recommendations to keep the animals healthy, and receive alerts when there are signs of sickness or stress, or when an animal might be pregnant.

Melisha's project is still in its infancy, but the early results have been promising. In the first field tests, the solution was about 95 percent accurate in predicting an animal's health. It already enabled one family to prevent a deadly outbreak by identifying a cow that was in the earliest stages of an anthrax infection, before symptoms were evident to the farmer.

Like Melisha's project, AI itself is still at a nascent stage. Thanks to advances in the past few years, we're beginning to build systems that can perceive, learn and reason, and on this basis, can make predictions or recommendations. Nearly every field of human endeavor could benefit from AI systems designed to complement human intelligence. From preventing once-deadly diseases, to enabling people with disabilities to participate more fully in society, to creating more sustainable ways to use the earth's scarce resources, AI promises a better future for all.

Change of this magnitude inevitably gives rise to societal issues. The computer era has required us to grapple with important questions about privacy, safety, security, fairness, inclusion, and the importance and value of human labor. All of these questions will take on particular importance as AI systems become more useful and are more widely deployed. To ensure that AI can deliver on its promise, we must find answers that embrace the full range of human hopes, needs, expectations and desires.

Conclusion
AI Amplifying Human Ingenuity

This will take a human-centered approach to AI that reflects timeless values. And it will take an approach that is firmly centered on harnessing the power of computational intelligence to help people. The idea isn't to replace people with machines, but to supplement human capabilities with the unmatched ability of AI to analyze huge amounts of data and find patterns that would otherwise be impossible to detect.

How AI will change our lives—and the lives of our children—is impossible to predict. But we can look to Melisha's device—a device that could help millions of small farmers in remote communities live more prosperously—to see one example of what can happen when human intelligence and imagination are augmented by the power of AI.

We believe there are millions of Melishas around the world—people young and old who have imaginative ideas for how to harness AI to address societal challenges. Imagine the insight that will be unleashed if we can give them all access to the tools and capabilities that AI offers. Imagine the problems they will solve and the innovations they will create.

This won't happen by itself. A human-centered approach can only be realized if researchers, policymakers, and leaders from government, business and civil society come together to develop a shared ethical framework for artificial intelligence. This in turn will help foster responsible development of AI systems that will engender trust. As we move forward, we look forward to working with people in all walks of life and every sector to develop and share best practices for building a foundation for human-centered AI that is trusted by all.

Addendum
China's Mission in the Future of AI

Addendum
China's Mission in the Future of AI

China is rapidly rising on the international AI stage with its influence increasing day by day. We see that more and more global AI experts are gathering in China. Whether from Microsoft or other companies, universities or scientific research institutions, AI experts are all rushing to participate in China's AI industry.

From a global perspective, China has unique advantages in advancing the technology and industrialization of AI in recent years since enjoying the largest user population of internet, mobile internet and smartphone in the world. According to a report published by the China Internet Network Information Center (CNNIC) in December 2017, the number of netizens in China reached 772 million with a penetration rate of 55.8 percent. That's 4.1 and 9.1 percent higher than the global average (51.7%) and the Asia average (46.7%), respectively. The report also shows that the number of mobile phone users in China has reached an impressive 753 million, with the proportion of internet users using mobile phones rose from 95.1 percent in 2016 to 97.5 percent in 2017.

We know that popularization and technological progression of AI depend on two major factors: the scale of available data and the number of users using the technology. China has unique advantages in both areas compared with other countries and regions in the world. According to CNNIC, the number of netizens in China had reached 731 million by the end of 2016, which is equivalent to the total population of Europe. The total volume of data generated by the internet and mobile internet has already given China a leading edge due to its scale. On November 11th, 2017 (China's "singles day" online shopping extravaganza) alone, more than one billion online payments with peak time of over 200 thousand TPS and about one billion delivery orders were registered.

The New Generation Artificial Intelligence Development Plan issued by the Chinese government clearly states that artificial intelligence will enable new technologies, new products, new industries, new business types, and new business models, and that it will trigger major changes in the economical structure, profoundly changing human production, lifestyles, and modes of thinking to trigger an overall leap of society productivity. Today, such changes are beginning to take place in the Chinese society. For example, mobile applications, such as toutiao.com, JD.com, and Tmall, are attracting the community's attention, dramatically changing or influencing the flow of basic social resources such as information, logistics, and capital. New business models, such as digital retail, smart manufacturing and digital finance, have emerged.

It is because the Chinese government attaches great importance to artificial intelligence that China has gained great opportunities in AI, as well as engendered great responsibilities for the development of the industry. On the one hand, China should contribute its achievements to the development of the global AI industry. According to the statistics of CNNIC, as of June 2017, the country had 592 AI enterprises, accounting for 23.3% of the world's total. In 2016, there were 30,115 AI-related patent applications in China. On the other hand, China also needs to consider the social responsibility of AI algorithms—how to avoid the negative influence of artificial intelligence on existing problems generated by redistribution and guidance of basic social resources, such as unfairness, injustice and lack of tolerance and how to direct AI to benefit societies by helping more people participate in development opportunities and helping each individual and organization achieve more by using AI.

Addendum
China's Mission in the Future of AI

As a Chinese saying goes: "good timing, geographical convenience and harmonious human relations contribute to people's success". From the central government to local cities and common people, China has already begun to fully embrace artificial intelligence. Today, it is the time when China needs to explore how to deepen trust in AI so that AI can play a greater role in China's society and economy. At the same time, the country should seize opportunities while shouldering its due responsibilities.

The Chinese government also has the willingness and inclination to strengthen its research on AI-related policies, regulations and ethics. The Next-Generation Artificial Intelligence Development Plan mentions the development environment of artificial intelligence as follows: by 2020, the country will initially establish AI-related ethical norms, policies and regulations in some areas; by 2025, the country will initially establish AI-related laws, regulations, ethical norms and policy systems, forming security assessment and management capabilities on AI systems; by 2030, the country will complete AI-related laws and regulations, ethical norms and the policy system.

As a global company, Microsoft encourages all parties to engage in extensive dialogue and continuous cooperation to actively develop human-oriented artificial intelligence with the aim of shaping a future for artificial intelligence that maximizes its potential and mitigates its risks. In establishing trust in artificial intelligence, China and the rest of the world are on the same starting line. Whether in digital transformation or rapid development of the Chinese economy globally, China has the willingness and ability to participate in development of the global digital economy, contribute China's wisdom, and assume due responsibilities. This is China's mission in the future of AI, as well as Microsoft's mission.

Digital Economy Brings Global Inclusive Opportunities

New technologies, such as cloud computing, big data, and artificial intelligence are activating a completely new form of economy: the digital economy. The China Academy of Information and Communications Technology(CAICT) believes that the digital economy, with digital knowledge and information as the key means of production and digital technology innovation at the core engine and modern networks as the primary carrier, is a series of economic activities that continuously improve the digitization and intelligence of traditional industries and accelerates the restructuring of economic development and governance through deep integration of digital technology and the real economy.

Today, the digital economy has become the new direction for China to transform its economic growth momentum. China is entering a new era of fully implementing digital transformation and vigorously developing the digital economy. In 2017, the annual report on the work of the Chinese government mentioned the term,"digital economy", for the first time, proposing that "the country will accelerate growth of the digital economy, allow enterprises and common people to benefit from it". As the world's second largest economy, digital transformation of the Chinese economy and the development of the digital economy will also have a profound impact on the global digital economy. China has nearly one-fifth of the world's population and will fully build a moderately prosperous society. Therefore, the development of the digital economy will help the 1/5 of the world's population benefit from the opportunities brought by digital economy.

Microsoft believes that China's digital transformation will not only bring more efficient and smarter productivity tools but will also bring inclusive development to every organization and every individual. Therefore, the country will achieve extraordinary results in this transformation and achieve a brighter future. A report released jointly by Microsoft and the market research firm IDC shows that by 2021, digital transformation will contribute to 1% of the annual growth of China's GDP, or 716 billion dollars. Every individual and organization has the opportunity to participate in the transformation through digital technology platforms and tools.

Artificial intelligence can be regarded as the engine of digital transformation and the digital economy. Especially as public cloud services that carry artificial intelligence become cheaper and cheaper, more regions are covered, and reliability and security become ever better and when people can now access artificial intelligence services through just a mobile phone. Artificial intelligence can be combined with edge computing and cloud computing to engineer new digital products and services. Through trained AI models, calculations can be made automatically near data sources, such as IoT devices, smart phones and smart industrial devices, in real time. Large-scale big-data batch processing can also be completed in the cloud through data centers provided sufficient bandwidth by networks. These automated AI computing can continuously turn data into insight or help provide smart services.

It is this automated, low-cost, ubiquitous AI computing that brings inclusive development opportunities to every organization and every individual in China.

Innovation in China, Innovation for the Whole World

While actively participating in the digital economy driven by artificial intelligence, China has also provided unique business scenarios and lab environment for global AI development. The academic research accomplishments of China's AI industry are also spreading all over the world through the Chinese AI community. The AI talent report from LinkedIn revealed that, as of the first quarter of 2017, there were more than 1.9 million professionals in the AI industry worldwide, 850,000 of whom were in the US and 50,000 in China. The report says that there were nearly 140,000 Chinese AI professionals worldwide.

Addendum
China's Mission in the Future of AI

Microsoft actively shares leading global technologies with China, one of the most important innovation centers worldwide, while gathering local Chinese wisdom to reciprocate the entire world. Microsoft Asia-Pacific R&D Group, Microsoft's largest R&D organization outside the United States with a large number of Chinese AI professionals, has achieved world-renowned results. Today, the research results from Microsoft Asia-Pacific R&D Group are spreading throughout the world through the company's platform.

Microsoft's "Xiaoice" is an artificial intelligence conversational chatbot technology born and developed in China. In 2014, Beijing-based Microsoft Search Technology Center Asia developed the first version of Xiaoice which was initially envisioned as a cute and adorable 16-year-old girl who can talk to humans on multiple social platforms. By creating the emotional computing framework, Xiaoice gradually formed a complete AI system focusing on emotional intelligence (EQ) in addition to IQ, through the integration of AI algorithms, cloud computing and big data, and by adopting a method of fast iteration.

The Future Computed

Dr. Harry Shum, Executive Vice President of Microsoft and head of Microsoft's AI and Research Division, advocated "the three principles of artificial intelligence creation", and Microsoft's Xiaoice is the best embodiment of these principles:

- Successful artificial intelligence must be a combination of IQ and EQ, more than IQ alone.
- Successful artificial intelligence must acquire independent intellectual property, rather than an intermediate state of technology;
- Successful artificial intelligence must augment human capabilities, rather than an alternative to human labor (such as products of "smart manufacturing" like industrial robot arms).

By understanding the context and semantics of dialogues, Xiaoice delivers a better AI experience than simple human-computer interaction, especially in demonstrating an AI emotional experience that embraces contemporary Chinese culture and contemporary Chinese internet culture. In May 2017, Xiaoice even launched an original poetry collection The Sunshine Lost Its Glass Windows. This was the first poetry collection completely created by artificial intelligence in human history. Xiaoice learned from the work of 519 modern Chinese poets born after 1920, and then gradually created its poems in its own style after more than 10,000 iterative studies.

In addition, Xiaoice's voice surpasses more than 98% of people in terms of elaborating on children's books. The test showed that Xiaoice spent 1/500 of time converting children's text books to audio format at 1/80000 of the cost compared with people at the same level. Today, audio books produced through Xiaoice on a large scale have already come to the market. On December 3, 2017, Xiaoice won the "2017 World Internet Leading Scientific Achievement Award". At the Boao Forum for Asia in April 2018, Xiaoice interacted with guests from all over the world, helping them experience first-hand Chinese AI products, which was highly recognized.

Addendum
China's Mission in the Future of AI

Today, Xiaoice has embarked on a path of globalization. Its global strategy is to establish a fully localized team in each country and region with a population of more than 100 million to ensure that Xiaoice is rooted in the native culture of the country. In 2015 and 2016, the Japanese version りんな, and the US version, Zo, were launched; in February and August 2017, the Indian version, Ruuh, and the Indonesian version, Rinna, were launched. As of August 2017, Xiaoice had more than 100 million users, with dialogue sessions exceeding 30 billion rounds. People can experience Xiaoice on mobile Internet platforms including Facebook Messenger, LINE, WeChat, QQ, Windows 10, Meipai, JD.com, Miliao, Mijia, Youku and more.

With Great Power Comes Great Responsibility

Thanks to artificial intelligence, China is now advancing at a faster pace in almost all areas. With such advancement come important questions:

How can artificial intelligence help eliminate the gap between rich and poor? Well-educated, technology-savvy people tend to see more benefits from technology, and most people, especially people without the benefit of higher education, may be unable to change the status quo of poverty on their own. Through AI technology, more accessible and universal AI products and services will be introduced to help change this situation. Especially in China, how can artificial intelligence improve the opportunities for low-education populations who live in rural areas, distant and border regions, etc. so that they can better participate in the new economy? In addition, artificial intelligence will replace some existing jobs and necessary skills but will also bring new opportunities in more technical positions. How can we help those displaced by technological changes to adapt and find rewarding work in this new environment? Ensuring their well-being will require government action through policies such as tax and benefits changes, ideas such as universal basic income (UBI) and more.

How do we ensure that AI technology protects the safety of people and property and poses no threat to human privacy and other basic rights? What kind of responsibility should product designers and technology owners have? For example, how do we ensure that self-driving cars do not get out of control, and who will bear responsibility for the consequences of such occurrences?

How do we prevent bias and assure fair competition rather than monopolies created by artificial intelligence? For example, content can be selectively recommended and filtered based on user preferences, potentially resulting in the reinforcement of cognitive prejudices in users, and creating a technology monopoly of their time and attention, potentially going so far when some technological products are as essential a basic resources as water or electricity. Some social applications have already become necessary for people's daily communication, monopolizing the relationships among people. Should China take control of similar technological monopolies and pay closer attention to technologies that may cause or exacerbate user prejudices?

In these areas of social responsibility around artificial intelligence, Microsoft believes that these are important questions that must be addressed through open and sincere dialogue. However, all effective solutions to these challenges will require a credible, reliable, and secure AI technology platform as a foundational element. This is the ultimate solution. Microsoft's Azure intelligent cloud platform operated by 21Vianet Group has been stewarded and optimized according to the legal, regulatory and compliance environment of China when it was first made available in the market. We also provide flexible interfaces and access methods to on-going accommodate new regulatory rules.

The AI Path with Chinese Characteristics

In view of AI's importance to China's digital transformation and development of digital economy, Microsoft and its partners have explored the "AI path with Chinese characteristics" following Microsoft's six values of artificial intelligence. Today, Microsoft's Intelligent Cloud Platform, based on Microsoft's six values, is fully driving China's digital transformation.

Addendum
China's Mission in the Future of AI

Digital Dock Builds "Central Nerve" for Future Shipping

As the world's largest heavy-duty equipment manufacturer, Shanghai Zhenhua Heavy Industry Co., Ltd. (ZPMC) has a significant position in the global shipping industry. At present, ZPMC holds 80% market share of the global port equipment, and its cranes and heavy-duty equipment are used by more than 250 ports in 97 countries. The company is transforming from a traditional manufacturer to a new-generation digital smart port service provider. Using Microsoft's Azure Cloud, the company built an IoT platform that connects devices, analyzes real-time data, and brings them together in a global monitoring center. Song Hailiang, Chairman of ZPMC, said: "We used to sell hardware, and now we sell software and services. On this path of change, ZPMC, as a leader in global port manufacturer and international company, collaborates with Microsoft to create new products, new services and new business models."

The Combination of Deep Learning and Spatio-temporal Data Allows City Managers to Predict Passenger Flow

In recent years, several stampede accidents occurred because of surges in passenger traffic in public areas, which has caused the government and society to pay attention to urban population flow control. This has driven researchers to try to predict passenger traffic in cities. Traditional prediction models were often based on predicting individual behavior, which created great inaccuracy in prediction on a large scale. This is due to the spatio-temporal nature of population flow data, and there are many independent factors. Microsoft

Research Asia developed a special network model for spatio-temporal data called deep residual-error network structure. This model divides the city into uniform grids and inputs population flow data (including mobile phones, taxi tracking, etc.) to calculate inflow and outflow of each grid square. It then simulates the time and spatial attributes and improves training accuracy to obtain more accurate results . In 2016, the Institute's traffic forecasting system was launched in Guiyang.

Addendum
China's Mission in the Future of AI

"Cloud" Makes Travel More Convenient

With the rise of the shared bicycle industry, major shared bicycle companies have accelerated their growth. As a star enterprise in the industry, Mobike has not only entered major cities in China, but also entered more than 130 cities around the world, becoming the world's largest intelligent shared bicycle platform. With the expansion of the business came surging pressures on data processing including managing the number of bicycles, users, usage, and management data. In response to this challenge, Mobike migrated its entire data platform to the Microsoft Azure Intelligent Cloud. With the flexible, secure and reliable Azure Intelligent Cloud Platform, the company now manages millions of bicycles and users worldwide. With the help of Azure IoT services, Dynamics and CRM solutions, as well as predictive analytics with Azure machine learning and cognitive services, Mobike can now predict when to move bicycles and the number of bicycles to be moved, and they can optimize routes so as to improve utilization of each bicycle and reduce operating costs through big data.

In an Upgrade to Smart Healthcare, Airdoc Enables More People to See the World Once Again

Every year in China, tens of thousands of people with diabetes lose their sight due to untimely treatment of diabetic retinopathy (DR). However, many diabetic patients take effective measures in time so they can avoid blindness. The key is early and accyrate diagnosis. As a growth company focusing on improving the cure rate of medical consultations through deep learning, Airdoc helps doctors improve efficiency and accuracy of diagnosis. In particular, the company helps doctors quickly conduct a complete screening and analysis of retinopathy to empower them to identify and diagnose DR and improve disease management. The company, by using the Microsoft Cognitive Toolkit, meets processing and real-time concurrent demands of large-scale medical data.

Smart Cloud Helps Implement RCESNIP

At the end of 2011, the State Council of the People's Republic of China launched the National Rural Compulsory Education Student Nutrition Improvement Program (RCESNIP), led by the Ministry of Education. The nationwide program has an annual budget of 18-20 billion RMB from the Central Government to cover 13,400 rural compulsory education schools from 699 designated impoverished rural regions in 29 provinces. It aims to help improve the nutritional status of rural students and promote education fairness for over 32 million students in the compulsory education grades. The China Development Research Foundation and Microsoft jointly built the "Sunshine School Meal" data platform based on Microsoft's Azure Intelligent Cloud platform. The platform uses the internet, big data and other technologies to monitor implementation of policies and provides objective and scientifically sound insight on the effects of implementation. By using data visualization, it provides a reliable data basis for policy recommendations. This data platform currently covers more than 10,000 schools. Since the implementation of "Sunshine School Meals" platform, the biggest beneficiaries have been children who have grown up healthily due to nutrition improvement despite living poverty. As the program is carried out extensively, more children will be able to eat nutritious food and grow healthier.

Help in STEM Education and Empower Imagination

Founded in 2013 in Shenzhen, Makeblock hopes to help more people enjoy the fun of creativity by reducing barriers. With the renewed interest and development in the STEM (Science, Technology, Engineering, and Mathematics) education market at home and abroad, the Makeblock team began to focus on integrating technology and education with an emphasis on STEM. They develop products suitable for the STEM education market including product lines that combine software, hardware, courses, and robotics events aimed at increasing interest in learning robotic programming and other knowledge among children. Today, Makeblock has over two million users in more than 140 countries around the world, and has established subsidiaries in the United States, the Netherlands and Japan. In its robotic and graphical programming products, Microsoft Cognitive Services provides artificial intelligence features such as face recognition and emotion recognition. The cognitive services enhance the company's brand awareness and influence. More schools and families have access to AI, and more children and inventors benefit from Microsoft's cognitive service and unleash imagination to the fullest.

Face Recognition Technology Helps Reunite Missing Children with their Families

Microsoft's Face Recognition API is a service based on Microsoft's Azure Intelligent Cloud that can compare the similarity of faces by scanning the face within images and using advanced algorithms to account for age and other variables. To help the charity organization "Baby Come Home" find missing children, applications based on Microsoft's Face Recognition API analyze 27 different facial features. Despite different shooting angles, facial expressions and aging, the application can accurately recognize similar facial images from many photos. By cooperating with "Baby Come Home", Microsoft is providing China's largest public tracing website with technical support for face recognition, and illuminating the path to home for tens of thousands of missing children.

Smart IoT Enables People with Visual Impairments to "Hear" Information about Bus

Tingting Bus is a mobile app that helps people with visual impairments travel quickly and easily around the public bus transit system in Guangzhou. In September 2016, as one of Guangdong's top ten projects to improve people's lives (public transportation category), seeing-eye devices were installed on 3,000 buses in Guangzhou. When visually impaired people want to take a bus, the app automatically informs them via audible prompts about the stops nearby and the relevant bus lines for their destination. It announces how many stops particular buses are away from the station when the rider shakes their phone. The implementation of this project has brought great convenience to people with visual impairments. The success of Tingting Bus cannot be separated from large-scale application of IoT sensors. With the increase of operating bus lines and users supported by the platform, the number of relevant devices and IoT sensors has also increased exponentially. The networking capabilities of Azure IoT Center and the Microsoft Azure Intelligent Cloud allows Tingting Bus' technology team to connect millions of IoT devices in an easy, reliable, and secure manner.

Addendum
China's Mission in the Future of AI

Conclusion

Microsoft's AI is widely empowering a new round of digital transformation based on artificial intelligence and big data. New technologies, such as artificial intelligence, virtual reality, mixed reality, big data, and cloud computing are creating a completing system, working together to enable end-to-end user experiences and solutions. Only by systematically and comprehensively pushing forward these new technologies in a responsible way can users ultimately come to trust in artificial intelligence, and realize the new digital economy to its full potential. In this way, China can better participate in the development of the global digital economy, contribute China's wisdom, and assume its due responsibilities.

The Future Computed

The best way to predict the future is to create the future. We believe that artificial intelligence will create a better future for mankind. Microsoft believes in and has confidence in such a future. Today, Microsoft and Microsoft Asia-Pacific R&D Group have sent a large number of technological experts to China and have promoted the development and global expansion of China's local economy through Microsoft's AI technology and intelligent cloud platform. Microsoft encourages open engagement and discussion around the challenges that new technologies will bring to China. And in establishing trust in these new technologies, Microsoft is willing to partner with China in the future of AI.

Notes

1. See Brad Smith and Carol Ann Browne, "Today in Technology: The Day the Horse Lost its Job," at https://www.linkedin.com/pulse/today-technology-day-horse-lost-its-job-brad-smith/.

2. Lendol Calder, Financing the American Dream: A Cultural History of Consumer Credit (Princeton: Princeton University Press, 1999), p. 184.

3. John Steele Gordon, An Empire of Wealth: The Epic History of American Economic Power (New York: HarperCollins Publishers, 2004), p. 299-300.

4. See Harry Shum blog, July 2017 at https://blogs.microsoft.com/blog/2017/07/12/microsofts-role-intersection-ai-people-society.

5. https://blogs.microsoft.com/ai/microsoft-researchers-win-imagenet-computer-vision-challenge.

6. https://www.microsoft.com/en-us/research/blog/microsoft-researchers-achieve-new-conversational-speech-recognition-milestone.

7. See Harry Shum blog, May, 2017 at https://blogs.microsoft.com/blog/2017/05/10/microsoft-build-2017-microsoft-ai-amplify-human-ingenuity.

8. https://www.microsoft.com/en-us/research/project/medical-image-analysis.

9. https://www.microsoft.com/en-us/research/project/project-premonition.

10. For example, when you ask Cortana "How big is Ireland?" the response is not only in square kilometers, but also says "about equal to the size of South Carolina."

11. https://www.microsoft.com/en-us/seeing-ai.

12. https://www.microsoft.com/en-us/research/project/farmbeats-iot-agriculture/#.

13. https://www.partnershiponai.org.

14. https://www.nytimes.com/2017/10/26/opinion/algorithm-compas-sen-

tencing-bias.html and https://www.propublica.org/article/machine-bias-risk-assessments-in-criminal-sentencing.

15. https://www.nytimes.com/2017/11/21/magazine/can-ai-be-taught-to-explain-itself.html.

16. Daniel Solove, "A Brief History of Information Privacy Law," [GW Law] 2006, p.1-25.

17. One interesting set of insights emerges from the transition from horses to automobiles. This gave birth to multiple new industries, many of which were impossible to predict when cars first came into use. https://www.linkedin.com/pulse/today-technology-day-horse-lost-its-job-brad-smith.

18. http://www3.weforum.org/docs/WEF_FOJ_Executive_Summary_Jobs.pdf.

19. http://query.nytimes.com/gst/abstract.html?res=9C03EEDF1F39E133A-25755C2A9649C946995D6CF&legacy=true.

20. https://www.economist.com/news/special-report/21700758-will-smarter-machines-cause-mass-unemployment-automation-and-anxiety.

21. https://www.economist.com/news/special-report/21700758-will-smarter-machines-cause-mass-unemployment-automation-and-anxiety.

22. https://www.economist.com/news/special-report/21700758-will-smarter-machines-cause-mass-unemployment-automation-and-anxiety.

23. https://venturebeat.com/2017/10/04/the-fundamental-differences-between-automation-and-ai.

24. https://www.washingtonpost.com/news/theworldpost/wp/2017/10/19/inside-chinas-quest-to-become-the-global-leader-in-ai/?utm_.term=.9da300d7d549.

25. **AI Survey. Risk Drivers.** https://news.microsoft.com/cloudforgood/.policy/briefing-papers/responsible-cloud/amplifying-human-ingenuity-artificial-intelligence.html.

26. https://www.oxfordmartin.ox.ac.uk/downloads/academic/The_Future_of_Employment.pdf.

27. https://openknowledge.worldbank.org/handle/10986/23347.

28. https://papers.ssrn.com/sol3/papers.cfm?abstract_id=2940245.

29. https://www.theguardian.com/technology/2017/jan/11/robots-jobs-employees-artificial-intelligence.

30. https://www.postandcourier.com/business/as-amazon-pushes-forward-with-robots-workers-find-new-roles/article_c5777048-97ca-11e7-955e-8f628022e7cc.html.

31. https://www.forrester.com/report/The+Future+Of+Jobs+2025+Working+Side+By+Side+With+Robots/-/E-RES119861.

32. https://www.economist.com/news/special-report/21700758-will-smarter-machines-cause-mass-unemployment-automation-and-anxiety.

33. "The new new way of working series: Twelve forces that will radically change how organizations work," BCG, March 2017. https://www.bcg.com/en-us/publications/2017/people-organization-strategy-twelve-forces-radically-change-organizations-work.aspx.

34. http://reports.weforum.org/future-of-jobs-2016/skills-stability/?doing_wp_cron=1514488681.1306788921356201171875.

35. https://www.technologyreview.com/s/515926/how-technology-is-destroying-jobs.

36. https://cew.georgetown.edu/wp-content/uploads/Americas-Divided-Recovery-web.pdf.

37. https://krueger.princeton.edu/sites/default/files/akrueger/files/katz_krueger_cws_-_march_29_20165.pdf.

38. http://www.oxfordmartin.ox.ac.uk/publications/view/1314.

39. http://www.hamiltonproject.org/papers/who_is_out_of_the_labor_force.

40. http://www.pewinternet.org/2016/11/17/gig-work-online-selling-and-home-sharing.

41. According to the Bureau of Labor Statistics, 6 million people are working part-time because that is their preference, an increase of 12 percent since 2007.http://www.bloomberg.com/news/articles/2015-08-18/why-6-million-americans-would-rather-work-part-time.

42. https://www.teacherspayteachers.com.

43. http://journals.sagepub.com/eprint/3FMTvCNPJ4SkhW9tgpWP/full.

44. http://globalworkplaceanalytics.com/resources/costs-benefits.

45. http://www.pewsocialtrends.org/2016/10/06/4-skills-and-training-needed-to-compete-in-todays-economy.

46. Furthermore, according to the National Center for Education Statistics, 1 in 5 high school students does not graduate within 4 years of beginning high school.

47. https://secure-media.collegeboard.org/digitalServices/pdf/research/2016/Program-Summary-Report-2016.pdf.

48. https://www.bls.gov/charts/job-openings-and-labor-turnover/opening-hire-seps-rates.htm.

49. https://www.bloomberg.com/news/articles/2017-06-22/the-world-s-workers-have-bigger-problems-than-a-robot-apocalypse.

50. https://www.nationalskillscoalition.org/resources/publications/2017-middle-skills-fact-sheets/file/United-States-MiddleSkills.pdf.

51. http://burning-glass.com/wp-content/uploads/2015/06/Digital_Skills_Gap.pdf.

52. https://www.nationalskillscoalition.org/resources/publications/file/Opportunity-Knocks-How-expanding-the-Work-Opportunity-Tax-Credit-could-grow-businesses-help-low-skill-workers-and-close-the-skills-gap.pdf.

53. The availability of broadband in remote and underserved communities can be instrumental in expanding the quality and accessibility of education, training and broader civic engagement. But there are 23.4 million people living in rural counties who don't have access to broadband and therefore do not have access to on-demand learning tools. To meet that need, in July 2017, Microsoft launched its Rural Airband Initiative to help serve as a catalyst for broader market adoption of this new model and to eliminate the rural broadband gap in the U.S. by July 4, 2022.
https://news.microsoft.com/rural-broadband.

54. One example of Microsoft's global skills initiatives is Microsoft India's Program Oorja, which works with polytechnics, industrial technology institutes and engineering colleges to enable students to be ready for work by helping them acquire certifications in various Microsoft Education curricula, largely in office productivity.
https://www.microsoft.com/en-in/about/citizenship/youthspark/youthsparkhub/programs/partners-in-learning.

55. https://news.microsoft.com/download/presskits/education/docs/IDC_101513.pdf.

56. https://news.microsoft.com/2017/06/27/the-markle-foundation-and-microsoft-partner-to-accelerate-a-skills-based-labor-market-for-the-digital-economy.

57. Just as more accurate and up-to-date data is needed to understand evolving jobs and needed skills, more data also is needed to better understand how employer and employee relationships and working conditions are evolving, including how the nature of work is changing. In addition, many existing government programs rely upon wage data to assess employment outcomes; a broader set of data may be needed to understand the true impact of newer contingent worker arrangements. Platform companies can contribute private-sector data to enhance this analysis.

58. Although online platforms, by most estimates, still only make up less than 1 percent of the workforce, the percentage of workers not in traditional employer/employee work arrangements (temporary agencies, on-call workers, contract workers, independent contractors or freelancers) is much greater. See, e.g., The Rise and Nature of Alternative Work Arrangements in The United States, 1995-2015.

59. In the absence of modernized laws, regulatory agencies are developing interpretations that represent vast departures from prior precedent -for example, expanding the scope of joint employment. With the changing political composition of many regulatory agencies, there is the potential for new case law that swings the pendulum in the opposite direction. The United States Congress is also proposing to legislate key definitions.

60. http://www.goodworkcode.org/about.

61. http://www.care.com.

62. See, e.g., S. 1251 and H.R.2685, Portable Benefits for Independent Workers Pilot Program Act, introduced by Senator Warner and Rep. DelBene. The act would establish a portable benefits pilot program at the U.S. Department of Labor, providing $20 million for competitive grants for states, local governments and nonprofits to pilot and evaluate new models or improve existing ones to offer portable benefits for contractors, temporary workers and self-employed workers.

63. We know from existing data that workers in recent decades already experience multiple instances of joblessness over a career. The National Longitudinal Survey of Youth 1979 (NLSY79) tracked a nationally representative sample of people born in the years 1957 to 1964; they experienced an average of 5.6 spells of unemployment from age 18 to age 48. High school dropouts experienced an average of 7.7 spells of unemployment from age 18 to age 48, while high school graduates experienced 5.4 spells and college graduates experienced 3.9 spells. In addition, nearly one-third of high school dropouts in the survey experienced 10 or more spells of unemployment, compared with 22 percent of high school graduates and 6 percent of college graduates.

> **我向诸君提出这一问题：机器能否思考？**

阿兰·图灵，1950年

第 1 章

人工智能
的未来

序
计算未来

早在20年前,我们二人就已经在微软共事,虽然那时我们分处地球的两端。那是在1998年,我们一个在中国,作为微软亚洲研究院的创始人成员,在北京的实验室里从早忙到晚;另一个则在5000英里之外的西雅图微软总部,负责领导国际法律及公司事务。我们生活在远隔重洋的两块大陆上,文化背景更是天差地别。虽然我们同在微软工作,但每天走进办公室之前的日常工作,却是迥然不同。

那时在美国,人们会在睡前打开咖啡机的定时开关,这样第二天一起床,就能立刻享有一杯香浓的咖啡——这可以算是技术自动化给生活带来的了不起的小成就。在享用咖啡的同时,美国人一般会看电视或看报纸来了解这个世界在你睡觉时都发生了什么大事。对于很多人来说,一本日记手札简直就是生命线,它会提醒自己在即将开始的一天中要做的所有事:从一大早办公室里的晨会、电话会议的接入号码和密码,到下午预约门诊的地址,以至于设置定时录像录制节目等,这些七七八八一长列待办事项。在出门上班之前,你可能已经打了一堆电话,比如叮嘱保姆该什么时间接送小孩,以及确认晚餐的安排等——即便对方没有接听也没关系,因为可以在自动答录机上留言。

同样是在20年前,对于大多数中国人来说,卧室里唯一的数字产品可能只有一台数字闹钟。人们习惯把事情写在台历上,例如在某一天约了什么人在哪儿见面,还有对方的电话号码。送孩子出门上学之后,人们会顺便从街边的早点铺带回油条和豆浆,然后一边吃早点,一边从收音机里了解世界动态。在1998年的北京,挤在拥挤的地铁和公交车上往返于市区内外的上班族们,常常会把脸埋在书报中抓紧时间阅读——而不像今天的"低头族"摆弄智能手机或电脑。

时过境迁,今天人们每天早上的基本活动其实并没有发生太大的变化,但在科技进步的驱使下,我们从事这些活动的具体方式却已经是今非昔比。如今,北京和西雅图的早晨仍旧有所不同,但它们之间的区别已经没有那么明显。发生在地球两端的情景大致如此:在床头柜上充电的手机把你叫醒,并为你推送当天的头条新闻和朋友圈里的最新动态;你会拿起手机,查看昨天晚上收到的电子邮件;发条信息给你的好友,确定晚餐安排;然后更新一条日历通知你的保姆,告诉她孩子足球训练课的具体时间地点;最后,在出门之前，查看今天的路况信息。今天，2018年，无论你在北京还是西雅图，都可以用智能手机点上一杯星巴克的双份脱脂拿铁，然后叫辆顺风车送你去上班。

仅仅是与20年前相比，我们今天司空见惯的很多东西都像是那时科幻小说的素材。毫无疑问，短短20年的改变，已经是翻天覆地。

那么，20年后的早晨又会变成怎样一幅情景呢？我们在微软勾画了这样一种可能：在你进入梦乡之后，你的专属个人数字助理小娜（Cortana）会着手整理你的全部日程，通过协调调度你家中的各种智能设备，她可以在你即将结束一个完整的睡眠循环时将你轻轻唤醒，不但保证让你以最轻松愉快的状态醒来，同时还能留出足够的时间让你冲个淋浴，穿戴整齐，出门上路，并准时出席今天的第一个会议。在你准备就绪之后，小娜会根据你的个人喜好和工作需要，为你阅读新闻、研究报告及社交媒体上的动态，通过梳理你的日历、会议安排、通信记录，以及正在做的项目和写下的文字，小娜能做到真正理解你的需求。此外，小娜还会为你更新各种信息和安排，例如天气变化、即将开始的会议信息、会面对象的情况介绍，还能根据实时路况提醒你应该何时出门。

参考你一年前的指令，小娜会记得这天是你姐姐的生日，并帮你预订一束她最喜欢的百合花，在当天下午送到她府上。当然，小娜也会把送花的事告诉你，这样在姐姐表示感谢时，你不会不知所措。如果需要，小娜还会在你们都喜欢的餐厅里预订座位，当然是在你们双方的行程安排都合适的时候。

2038年，数字设备将帮助我们真正用好我们最宝贵的财产——时间。

20年后，不出家门就能参加工作会议，只要戴上HoloLens或类似设备，你就可以借助混合现实技术，在虚拟会议室里与同事和客户进行身临其境的交谈和互动。即使语言不通也不用担心，你的演示和谈话都会被自动翻译成不同语言，每一位与会者都能通过耳机或电话听到他们母语的实时翻译。小娜这样的数字助理会自动制作一份会议纪要，不但能根据会上进行的对话和决策，列出每位与会者接下来分配的任务，还能在与会者的日程表上自动添加相关工作的日程提醒。

2038年，在无人驾驶汽车送你去参加会议的途中，可以利用车载数字终端继续完善自己的演讲。小娜将帮你从最新发表的文章和报告中提取研究成果与数据，将最新信息绘制成一目了然的图表，供你参阅。根据你的指令，小娜可以自动回复常规电子邮件，或者将可由其他人员处理的邮件转发出去，在发送时，还会根据项目时间表，要求相关人员在一定时间内作出回复。实际上，上述有些功能在今天已经实现,但20年后,所有这些功能都将变成每个人司空见惯的日常操作。

还有越来越多的智能设备帮你监控自己的健康状况。如果发现了任何异常征兆,小娜就会帮你预约医生,她还会帮你长期记录和定期安排常规体检、疫苗接种和免疫测试。数字助理预约和安排就医时,会确保选择你最方便的时间。结束工作后,无人驾驶汽车将带你回家,然后在家里通过远程连线接受医生的虚拟检查。各类移动设备为你测量血压、血氧,并将结果传送给医生,医生在通话过程中就能对这些数据进行分析和诊断。通过比对超过TB级别的海量健康数据,人工智能可以协助医生对检查结果进行准确的分析和诊断,根据你个人的生理特征为你量身定制治疗方案。几个小时内,无人机将把药物送到家门前,由小娜来提醒你按时服药。小娜会持续监控你的康复进展,如果病情未见改善,她会在征得你的同意后,为你预约复诊。

在自动化的未来,当你想要放松自我的时候,不必再去联系旅行社或上网预订机票和酒店,只要动动嘴说:"嗨,小娜,请为我制订一个两周的休假计划。"小娜会根据季节、预算、你的时间和兴趣,帮你量身打造一份行程,而你要做的就是决定目的地,然后动身上路。

回首过去,20年间科技的进步,为我们工作和生活的方式带来了翻天覆地的巨变。由云计算驱动的数字技术，让人类变得更具智慧，也让我们能够更高效地利用时间，更有效地提高生产力，并且更顺畅地相互交流。而这，还仅仅是个开始。

不久的将来，许多枯燥的重复性劳动将由人工智能自动处理，让我们将宝贵的时间和精力用于更具建设性和创造性的工作。推而广之，人工智能将让人类有能力驾驭真正的海量数据，并帮助我们在诸如医疗、农业、教育、交通等领域取得突破性的进展。事实上，今天我们已经看到，人工智能有能力帮助医生减少医疗事故、帮助农民增加收成、帮助教师因材施教、帮助科研人员为保护地球家园找到更好的解决方案。

与此同时，过去20年来的经验也告诫我们：数字技术的进步带来的不仅有便利的生活，还会引发许多复杂的问题，而科学技术对人类社会的潜在影响，总能引起人们最广泛的担忧。随着互联网时代的到来，技术已经成为我们工作生活中不可或缺的一部分，今天我们在饭桌边讨论智能手机为什么能让人们终日魂不守舍，同时，人们也在公开探讨诸如网络安全、个人隐私和社交媒体被恐怖主义活动利用的隐忧。这样的广泛关注不但促成了一系列全新的公共政策与法律法规的出台，也催生出全新的法律研究领域，并引发了人们对于计算机科学领域中伦理道德问题的思考。随着人工智能的不断进化，以及全世界对于人工智能将要扮演社会角色的日渐关注，这样的讨论无疑将会持续下去。放眼未来，我们既要保持开放的态度,也要保留质疑的精神,只有这样,我们才能成竹在胸地面对人工智能为我们带来的机遇与挑战。

透过过去20年间隐私保护条款不断完善的过程,我们或许可以窥见未来数年围绕人工智能将会出现的一些趋势和变化。1998年,即使在美国也很难找到一位全职的"隐私律师"——在全球首批数字隐私法律中,最知名的《欧洲共同体数据保护指令》(European Community's Data Protection Directive)是在1995年通过的,而这一领域的专业领导机构"隐私专业人员国际协会"(International Association of Privacy Professionals,IAPP)在2000年才成立。

如今,IAPP已经拥有来自全球83个国家的2万多名成员,每次举办会议都有近万名听众与会。IAPP成员讨论的议题内容非常广泛,既包括企业责任,也有与用户信息搜集、使用、保护相关的道德伦理问题。如今的隐私律师也不会缺少案源,已有100多个国家和地区设立了数据保护机构,来负责隐私监管问题。20年前,隐私监管作为独立的法律分支学科几乎还不存在,如今已一跃成为举足轻重的法律领域之一。

未来,涉及人工智能的议题、政策和法规会产生哪些变化?在计算机科学领域,对人工智能的种种疑虑,是否意味着计算机程序员和研究人员必须学习伦理学?我们认为答案可能是肯定的。那么,未来的程序员是否也要如同医生宣誓

《希波克拉底誓言》一般遵从某种道德准则？这层考虑不无道理。我们要共同学习，并坚决履行我们的社会责任。最终的问题不是计算机能够做什么，而是计算机应当做什么。

那么，未来是不是会出现"人工智能法"这一全新的法律领域呢？现在人工智能相关法律的发展就像数据隐私法在1998年面对的情况：部分现有法律可应用于人工智能领域，特别是侵权法和隐私法，与此同时，也将出现一些新的具体法规，如无人驾驶汽车相关法规。但人工智能法尚未成为独立的法律领域，现在我们去参加会议时，肯定不会有人自称是"人工智能律师"。但到2038年，情况几乎可以肯定会有所改变，那时，不仅会出现专门处理人工智能诉讼的人工智能律师，而且几乎所有律师，都将借助人工智能辅助执业。

真正的问题并非"人工智能法"是否会出现，而是在什么时间、以何种方式，才能最好地制定出这样的法律。目前我们对此还没有答案，但是非常幸运的是，每天与我们一起工作的人们，已经提出了正确的问题。正如他们所言，只有在人工智能科技不断发展成熟之后，才能制定出有针对性的监管规则。在那之前，我们首先需要形成一套统一的社会原则和价值观，用以管理人工智能的发展和应用，在此基础上，我们要按照这些社会原则和价值观开发出最佳实践经验。在这之后才是政府出面的更好时机，由政府参照之前的实践经验，来制定法律和监管条例，要求每个人都按章行事。

完成这些工作需要假以时日，也许要好几年，但一定不会超过20年。目前，我们或许可以将六大道德原则作为起点，以指导人工智能的发展与利用。这六大道德原则将确保人工智能体系公平、可靠与安全、隐私与保障、包容、透明、负责。我们对这些原则或类似原则了解越多，技术开发人员和用户就越能通过最佳实践执行这些原则，我们便能为人工智能监管制定出适当的社会规则，从而更好地服务整个世界。

目前，有些人认为，在推动人工智能发展的过程中，我们只需要道德准则和最佳实践。他们相信，科技创新本身并不需要监管机构、立法者和律师的帮助。

虽然他们阐述了一些重要的观点，但我们认为这样的想法是不切实际的，甚至可能产生误导。与古往今来的所有科技成果一样，人工智能必然会造福于人类社会，但也不可避免地，可能被一些不法分子所利用。正如不法分子会通过邮局进行邮件欺诈、利用电信服务进行电信诈骗一样，自1998年以来，互联网在推动社会进步的同时，也成了欺诈犯罪角逐的舞台，犯罪手法层出不穷、愈演愈烈，波及范围也遍布全球。

我们不得不设想，到2038年，犯罪集团或其他组织很可能会通过滥用人工智能技术，危害社会。与此同时，围绕着如何以社会可以接受的方式利用人工智能的话题，也一定会涌现出很多棘手难题。要有效地解决和处理这些问题，都要依赖于制定全新的法律法规。因此，我们既不能在尚未弄懂将要面临的问题之前，轻率地制定法律，阻碍人工智能技术的发展；与此同时，我们也不能无所作为，干等到20年后再着手制定相关法律——我们需要在这二者之间找到一个平衡点。

在考虑制定人工智能监管的原则、政策和法律时，我们还必须关注人工智能将对全世界就业造成的影响。哪些工作岗位会因人工智能而消失？人工智能又将创造出哪些新的工作岗位？过去250多年来，在人类科技发展史上有一件事是恒久不变的，那就是科技进步对就业的影响——创造新的工作、消灭既有工作、颠覆工作的职责和内容——今天人工智能的发展，也不会避开这一规律。

人工智能会创造更多的工作岗位，还是取代更多的工作岗位？经济学家指出，之前每次工业革命所创造的工作岗位的数量，都会超过被取代的工作岗位的数量。我们有充分的理由相信，这种情况也会出现在人工智能的发展过程中，但未来究竟怎样，目前还难以作出定论。

我们很难准确预测就业市场的未来走向，因为科技发展对就业的影响往往是间接的，而且会受制于诸多相互作用的创新和事件。拿汽车取代马车的历史进程举例来说，随着汽车的普及，制造马车的工作岗位无疑变得越来越少，但与此同时，在生产汽车轮胎的工厂里又出现了更多全新的工作岗位。而这只是汽车所带来影响的冰山一角而已。[1]

汽车的兴起首先引发了20世纪20年代末到30年代美国农业的大萧条，全美的经济都因此受到波及。原因何在？因为在马匹数量迅速下降后，美国农民的收入也随之减少。在汽车盛行之前的10年，美国约四分之一的农产品被用来喂马，马匹数量减少意味着干草需求量大幅降低，农民不得不转而种植其他农作物。当市场因此充斥着这些农产品，就又造成了农产品价格的大幅下滑。农业的萧条冲击了农村的地方银行，进而影响了整个金融体系。

但在另一方面，汽车的兴起也对经济发展造成了间接的积极影响：随着汽车销量的增加，一些乍看与汽车风马牛不相及的行业部门也发展起来，其中之一便是新兴的消费信贷行业。亨利·福特发明的流水生产线，让汽车成为多数家庭可以负担的商品，但由于汽车价格依然高昂，大家不得不先行借款。正如一位历史学家所言："分期还款信贷行业的繁荣与汽车行业的成功互为因果。"[2]简而言之，汽车引爆了一个全新的金融服务市场兴起。

广告业同样受益于汽车行业的发展。在汽车行驶速度达到每小时30英里以上时，"只有能够被一眼辨识的标志能够被驾驶员看到，否则将被忽略掉"[3]。这就是公司品牌标志产生的原因之一——无论公司标志树立在哪儿，都应该被一眼认出。

我们再来看一下汽车对纽约曼哈顿岛造成的间接影响。在百老汇畅行的汽车，为华尔街的金融业创造了新的工作机会，在麦迪逊大街，则为广告业带来了更多的工作岗位。但当汽车首次亮相于城市街道时，几乎没有人预见到汽车会创造出这些新的工作岗位。

基于这些过往的经验，我们需要对人工智能及其他未来技术可能对就业产生的影响保持警醒。虽然我们大体上已经可以预言，新的工作岗位会出现，一些既有的工作岗位会消失，但我们始终都要坚信，无论如何，我们都绝对有能力适应任何可能到来的意外情况。

在准备迎接未来的不确定性时，有一件事是毋庸置疑的：新的工作岗位要求新的技能。事实上，很多旧的工作岗位也会要求新的工作技能。这是科技发展的必然要求。

让我们回顾一下过去30年间所发生的变化。现在，每个中等规模以上的企业机构都会有一名或多名专职的IT支持人员，而在30年前，这类工作岗位几乎不存在。而且，并不是只有IT工作人员才需要掌握IT技术。20世纪80年代初，办公人员一般都是先手写公文，然后交由秘书用打字机打出来。到了80年代末，秘书们学会了使用文字处理机。到了20世纪90年代，因为所有办公人员都学会了用电脑制作公文，秘书人数随之急剧下降。今天，我们都知道，IT培训并不只是针对IT专业人士的。

同样，对掌握数字技术和其他新技能的专业人士的需求也在不断增长，有些部门严重缺乏此类专才。在我们迎来第四次工业革命时，不仅是编程和计算机科学，其他地位逐渐提升的重要领域，如数据科学，同样需要大量技术人才。今天的重点，已不再是鼓励大家学习新技能，而是帮助大家找到学习技能的新途径。调查显示，父母们特别希望自己的孩子能够有机会学习编程。在微软内部，当我们为员工开设与人工智能最新进展相关的课程时，大家的响应总是异常火爆。

当前最大的挑战是如何创造出更多的途径帮助大家学习新技能，并让劳动力市场的运作方式与之适应，以便帮助劳资双方更灵活地找到自己的新位置。许多社群和市场，已经作出了各种创新来应对这一挑战，我们可以从中汲取很多经验。其中有些创新是为早已存在的项目提供了新的实施途径，例如在瑞士大获成功的青年学徒制职业教育模式。还有一些新近推出的创新，包括领英（LinkedIn）等公司推出的网络工具和服务，以及非营利机构（如马克尔基金会）在科罗拉多州实施的"熟练技术"项目等。

人工智能、云计算以及其他新科技带来的影响与改变远不止于此。几十年前，大多数国家普遍存在的雇佣关系都非常传统，受雇方的工作地点要么是在办公室，要么是在制造工厂。科技的发展颠覆了这一模式,许多人开始作为合同工从事远程或兼职工作,或提供基于项目的合作。大多数研究均认为,这种趋势在未来将不断壮大。

为了让人工智能以及其他科技能够尽可能地为社会创造更多福祉,我们需要修订劳动法律和相关政策,以应对新的形势。多数现行劳动法律法规是为了适应20世纪初的科技创新而制定的,但在一个世纪之后的今天,这些法律法规已经不再符合雇主和雇员的需求。例如,大多数国家的劳动法均假设,所有雇员要么是全职雇员,要么是独立合同工,而完全没有考虑到那些服务于优步(Uber)、Lyft等新兴行业的人群;这些新兴行业覆盖了诸多领域,包括技术支持和看护服务。

健康保险和其他福利计划也存在类似情况,因为这些保险和福利计划的设计前提是全职雇员多年受雇于同一雇主;这种设计不适用于同时服务多家公司或经常更换工作的人群。我们的社会保障网(包括美国的社会保障制度)是20世纪上半叶的产物,所以,这些重要的公共政策亟须修订,以适应不断变化的世界。

在展望未来时,世界变化的速度,或许会令我们望而生畏;但回想1998年以来科技的发展,我们应该感到欣慰,因为我们成功应对了其中的挑战。展望2038年,我们可以预见到,未来即将发生巨变——这些巨变将会在全世界的各个角落创造无数机遇,同时也将带来无尽的挑战。

我们得到的主要结论

首先,在人工智能时代,只有迅速、有效地拥抱时代变化的企业和国家才能获得成功。原因很简单:人工智能发挥作用的领域,也正是人类智力发挥作用的领域,在几乎所有这些领域中,人工智能都会帮助我们提高生产力,实现经济增长。简言之,只有善用新科技者而非拒绝新科技者,才能实现工作机会和经济的双增长。

其次,虽然我们确信,人工智能将提升我们的生活品质、帮助解决重大社会问题,但与此同时,我们务必保持清醒的头脑。人工智能所带来的既有机遇也有挑战。这就是为什么我们不能仅仅关注科技发展本身,同时还应当关注其他重要议题,如确立有效的道德准则、修订法律法规、培训新技术人员,甚至实施劳动力市场改革。如此,我们才能最大化地利用人工智能新科技。

再次,我们应携手应对相关挑战,共同为我们的社会负起责任。原因之一是人工智能技术并非仅仅关乎科技行业。微软努力的方向是"普及人工智能全民化",正如同我们曾经致力于推广"电脑普及化"一样。20世纪70年代,我们曾成功地帮助各组织机构实现了对电脑的自定义应用，未来，我们也会帮助他们实现对人工智能技术的自定义应用。我们的策略是向每一人和每一组织提供最基础的人工智能工具，如计算机视觉、语音、知识认知服务，让他们能够设计出自己的人工智能解决方案。我们认为，这种做法要远胜于仅仅将人工智能的未来交由少数几家公司掌控。只有让更多的人有机会创造人工智能体系，才能让更多的人担负起解决人工智能相关问题与挑战的共同责任。

在科技飞速发展的今天，从事人工智能、云计算和其他创新技术开发的人员，最为了解科技的工作原理，但这并不意味着我们最懂得应当如何在社会生活中应用这些科技。因此，我们需要与各方携手，共同规划未来——包括政府机构、学界、商界、公民社会和其他利益相关方。这已经不再是单个社群或某一国家的任务，而正逐渐成为全世界共同的使命。我们每个人都有责任参与其中，并发挥我们不可或缺的一份力量。

所有这些都让我们得出一个极为重要的结论，就像史蒂夫·乔布斯生前反复强调的——要让工程技术与人文科学相互结合。

我们两个人，一个从小钻研计算机科学，另一个则一直从事人文研究。在微软共事多年后，我们两人的共识是——

在未来，这二者的结合将愈加重要。

一方面，人工智能的发展需要更多人掌握数字技术和数据科学；与此同时，在由人工智能驱动的未来世界，人们所需要具备的专业技术绝不限于科学、科技、工程和数学领域。当电脑能够高度模仿人类时，社会学科和人文学科将变得尤为重要。语言、艺术、历史、经济学、伦理学、哲学、心理学和人类发展学科所教授的批判性思维、哲学和伦理分析能力，对人工智能应用的发展与管理同样至关重要。要让人工智能最大限度地发挥其潜力，为人类服务，每个工程师都应该更深入地了解人文科学，而每个人文学科的学生也都应该更深入地了解工程学知识。

我们需要用更多的时间来彼此交谈、彼此倾听、彼此学习。我们两人虽然来自不同领域，但都从相互交流中获益良多——这一过程不仅珍贵，同时也愉悦无比。

在即将驶入新时代之际，希望本书能对您有所帮助。

致谢

衷心感谢以下各位为本书撰写提供的珍贵的见地和视角：

Benedikt Abendroth, Geff Brown, Carol Ann Browne, Dominic Carr, Pablo Chavez, Steve Clayton, Amy Colando, Jane Broom Davidson, Mariko Davidson, Paul Estes, John Galligan, Sue Glueck, Cristin Goodwin, Mary Gray, David Heiner, Merisa Heu-Weller, Eric Horvitz, Teresa Hutson, Nicole Isaac, Lucas Joppa, Aaron Kleiner, Allyson Knox, Cornelia Kutterer, Jenny Lay-Flurrie, Andrew Marshall, Anne Nergaard, Carolyn Nguyen, Barbara Olagaray, Michael Philips, Brent Sanders, Mary Snapp, Dev Stahlkopf, Steve Sweetman, Lisa Tanzi, Ana White, Joe Whittinghill, Joshua Winter, Portia Wu。